Four Philosophers and the Bomb

In this book, Alberto Castelli, Giunia Gatta, Micaela Latini, and Francesco Raschi examine how four prominent intellectuals of the 20th century (Bertrand Russell, Karl Jaspers, Raymond Aron, and Günther Anders) understood atomic warfare. With a chapter devoted to the philosophical ideas of each thinker and how they understood and interpreted war, the authors analyze the historic-political context in which these ideas emerged and what they proposed to avoid a nuclear disaster.

Four Philosophers and the Bomb will be of interest to students and researchers of peace studies, international relations, political philosophy, and moral philosophy.

Alberto Castelli teaches History of Political Thought at the University of Insubria. Castelli's research interests primarily concern the critique of violence as a means to change (for the better) the world.

Giunia Gatta teaches political philosophy at Bocconi University. She is interested in those aspects of politics that defy rationalization and in the role of emotions in motivating political action. The authors who most inform her thinking are Karl Jaspers, Judith Shklar, and Hannah Arendt. Her work has appeared in *Contemporary Political Theory*, *Philosophy and Social Criticism*, *The Review of Politics*, among others, and her book on Shklar was published by Routledge in 2018.

Micaela Latini is Associate Professor of German Literature at the University of Ferrara, where she also teaches Aesthetics. She is a member of the Doctoral Program in Human Sciences at the University of Ferrara. Her research is located at the intersection of German literature, philosophy, animal studies, and visual studies. Her current focus is on the images of contemporary wars and representations of animals in German literature and culture from the 18th to the 21st century.

Francesco Raschi is Associate Professor of History of Political Thought at the Department of Political and Social Sciences, University of Bologna. His teaching activity takes place at the Forlì Campus. His research interests include post-revolutionary political liberalism and contemporary liberalism.

Routledge Studies in Social and Political Thought

This series explores core issues in political philosophy and social theory. Addressing theoretical subjects of both historical and contemporary relevance, the series has broad appeal across the social sciences. Contributions include new studies of major thinkers, key debates and critical concepts. The full series can be viewed here.

Being a (Lived) Body
Aesthesiological and Phenomenological Paths
Tonino Griffero

Revisiting Social Theory
Challenges and Possibilities
Edited by D.V. Kumar

Alfred Schutz, Phenomenology, and the Renewal of Interpretive Social Science
Besnik Pula

Communicative Reason
A Sociological Restatement
Patrick O'Mahony

Myth, Society and Profanation
William Pawlett

Creating Democracy
Hannah Arendt and Mikhail Bakhtin
Charles Hersch

Four Philosophers and the Bomb
Russell, Aron, Jaspers and Anders on Atomic Warfare
Alberto Castelli, Giunia Gatta, Micaela Latini and Francesco Raschi

Four Philosophers and the Bomb
Russell, Aron, Jaspers, and Anders on Atomic Warfare

Alberto Castelli, Giunia Gatta, Micaela Latini, and Francesco Raschi

Routledge
Taylor & Francis Group
NEW YORK AND LONDON

First published 2025
by Routledge
605 Third Avenue, New York, NY 10158

and by Routledge
4 Park Square, Milton Park, Abingdon, Oxon, OX14 4RN

Routledge is an imprint of the Taylor & Francis Group, an informa business

© 2025 Alberto Castelli, Giunia Gatta, Micaela Latini and Francesco Raschi

The right of Alberto Castelli, Giunia Gatta, Micaela Latini and Francesco Raschi to be identified as authors of this work has been asserted in accordance with sections 77 and 78 of the Copyright, Designs and Patents Act 1988.

All rights reserved. No part of this book may be reprinted or reproduced or utilised in any form or by any electronic, mechanical, or other means, now known or hereafter invented, including photocopying and recording, or in any information storage or retrieval system, without permission in writing from the publishers.

Trademark notice: Product or corporate names may be trademarks or registered trademarks, and are used only for identification and explanation without intent to infringe.

Library of Congress Cataloging-in-Publication Data
Names: Castelli, Alberto, 1967- author. | Gatta, Giunia, 1971- author. | Latini, Micaela, 1973- author. | Raschi, Francesco, author.
Title: Four philosophers and the bomb: Russell, Aron, Jaspers and Anders on atomic warfare / Alberto Castelli, Giunia Gatta, Micaela Latini and Francesco Raschi.
Other titles: Russell, Aron, Jaspers and Anders on atomic warfare
Description: New York, NY: Routledge, 2025. |
Series: Routledge studies in social and political thought |
Includes bibliographical references and index. |
Identifiers: LCCN 2024057079 (print) | LCCN 2024057080 (ebook) |
ISBN 9781032966267 (hbk) | ISBN 9781003617174 (ebk)
Subjects: LCSH: Nuclear warfare–Philosophy. | Nuclear warfare–Moral and ethical aspects. | Russell, Bertrand, 1872–1970. | Jaspers, Karl, 1883–1969. | Aron, Raymond, 1905–1983. | Anders, Günther, 1902–1992. | World politics–20th century.
Classification: LCC U263 .C38 2025 (print) | LCC U263 (ebook) |
DDC 355.02/1701–dc23/eng/20250117
LC record available at https://lccn.loc.gov/2024057079
LC ebook record available at https://lccn.loc.gov/2024057080

ISBN: 978-1-032-96626-7 (hbk)
ISBN: 978-1-041-02000-4 (pbk)
ISBN: 978-1-003-61717-4 (ebk)

DOI: 10.4324/9781003617174

Typeset in Times New Roman
by Deanta Global Publishing Services, Chennai, India

Contents

Introduction   1
ALBERTO CASTELLI

Notes 5
Bibliography 5

1   Bertrand Russell's Commitment against Atomic Warfare   6
ALBERTO CASTELLI

Bertrand Russell and War 6
The Shock of the Atomic Bomb 7
The Need for a Change 8
World Government and Western Values 9
The H-Bomb and Its Enemies 11
"Remember Your Humanity and Forget the Rest" 12
The Intellectuals' Engagement 14
Overcoming Fanaticism and Nationalism 16
The Future of Man 18
The Strength of Reason and the Reasons of Strength 20
Notes 21
Bibliography 24

2   Raymond Aron: International Relations in the Atomic Age   26
FRANCESCO RASCHI

Introduction 26
Impossible Peace and Improbable War 28
Solidarity between the United States and the Soviet Union in
the Age of Deterrence 30

Massive Reprisals and Flexible Response 34
Conclusions 38
Notes 40
Bibliography 42

3   Karl Jaspers: Between the Bomb and Totalitarianism　　44
GIUNIA GATTA

Jaspers' Uneasy Relationship with Politics 47
The Future for Mankind, between Hard Realities and Lofty
Aspirations 49
Totalitarianism and the Bomb 51
What Is to Be Done? 54
Reason and Ideology 56
Notes 58
Bibliography 61

4   Hiroshima Is Everywhere: Günther Anders' Reflection on
     the Atomic Threat　　63
MICAELA LATINI

From the Human Being without a World to the World without
a Human Being 63
The Beginning of the End Times 67
Apocalyptic Blindness 73
To Be and Not to Be 76
Hiroshima Is Everywhere 78
Notes 81
Bibliography 83

Index　　85

Introduction

Alberto Castelli

This book is about the ideas of Bertrand Russell, Karl Jaspers, Raymond Aron, and Günther Anders on atomic warfare. Most of these ideas were formulated between the second half of the 1940s and the early 1960s. This period was marked by tensions due to the progressive intensification of the arms race, the Korean War, the invasion of Hungary, and the Suez Crisis, and characterized by the feeling of being on the brink of a third world war, in comparison to which the first two would have seemed trivial. In short, Russell, Jaspers, Aron, and Anders found themselves reflecting in a gloomy, frightening atmosphere, marked by the fear that the race towards self-destruction might be unstoppable. Our own time bears some resemblance to that period: we find ourselves in the midst of strong international tensions, and it would be difficult to argue that the atmosphere in which we live can support any form of carefree confidence in the future. Yet, our world is also different from that of the authors examined in this book: today, there is no longer a direct and constant confrontation between two superpowers threatening each other with atomic weapons, and I hope I am not soon proven wrong when I say that a new world war between major powers is unlikely. Why, then, revisit the reflections of Russell, Jaspers, Aron, and Anders today? First of all, for the obvious reason that every philosophical idea from the past has something to teach us; but also, because, although the era of the four authors we are dealing with does not resemble our own, it is nonetheless true that the problem they courageously and profoundly addressed – the possibility of nuclear holocaust – has not disappeared at all.

Indeed, the willingness to use nuclear weapons has hardly diminished over the eighty years that separate us from August 1945. Geopolitical contexts have changed, but those who have held, and still hold, the fate of the world in their hands have never abandoned the logic of the will to power, the principle "right or wrong, my country," or the older one, "si vis pacem para bellum" (if you want peace, prepare for war). Even today, that logic and those principles are considered obvious and accepted, with complete oblivion to the fact that "power, beyond a certain limit, turns into its opposite," as it annihilates everything for which it was sought in the first place.[1] Today, there are approximately 12,100 nuclear warheads scattered around the world, 2,100 of which are ready to be used in very short order. In the last 20 years, moreover, the expenses for the maintenance and modernization of arsenals have consistently increased (in 2023, approximately 91 billion dollars were spent overall).[2] Thus, despite the horror of the atomic bombings of Hiroshima and Nagasaki, the use of nuclear weapons remains a viable option for the governments of the nine countries in the "nuclear club" (China, France, India, Israel, North Korea, Pakistan, Russia, the United Kingdom, and the United States).

We can say that, while nuclear weapons ceased to frighten us after the end of the Cold War, it is not because their threat had disappeared, but because we had almost forgotten about them. Instead of developing an atomic consciousness, recognizing that "peace is not an inevitable process but a conquest (and like all conquests, it can be lost once achieved)," we have formed a kind of atomic unconsciousness, indulging in a pleasant collective amnesia in the face of the possibility of catastrophe.[3]

It is likely that this atomic unconsciousness has not been unwelcome to the states that possess nuclear weapons; indeed, some have even put forth the thesis that the issue of atomic bombs should no longer be regarded as urgent because, with the end of the confrontation between the United States and the USSR, the countries that possess them are responsible and do not wish to wage war on anyone. This is a specious argument: as I write, two of these responsible countries are engaged in bloody wars with no end in sight, and escalation is not out of the question.

The oblivion surrounding the nuclear issue has been influenced by at least two other factors: the first is that, starting in the early 1990s, Western public opinions have become accustomed to wars occurring near their borders. These were bloody but limited wars that did not directly and massively impact European or American citizens and whose consequences were not perceived by them as too burdensome.[4] In short, once the fear of the communist superpower disappeared, the last 30 years have taught Westerners that wars, even when they break out at their doorstep, should not cause too much concern about the possibility of the use of nuclear weapons. The second factor that has contributed to the oblivion of the atomic issue is that, in recent decades, other extremely urgent global problems have emerged, concerning the economy, the environment, rights, health, and new information technologies – global problems that have intertwined with other issues of regional or macro-regional significance, such as immigration, inflation, the energy crisis, and the instability of representative political systems. Neither the people nor their leaders tolerate a list of concerns that is too long.

It took the invasion of Ukraine by the Russian superpower (along with repeated threats to use nuclear weapons), followed by the indirect intervention of NATO, to change things and bring about a partial "return of the repressed." Partial, I wrote, because even in the three years since February 24, 2022, despite the fear of Russian weapons thundering at the gates of Europe, a certain indifference has persisted among citizens and government elites regarding the possibility of a nuclear escalation. People continue to believe – rightly or wrongly – that even though both sides involved in this war possess nuclear weapons, nothing more horrifying than what has happened in the Balkans, Chechnya, Iraq, Afghanistan, and elsewhere over the past 30y years will occur, and that, in short, Westerners will not pay too high a price for this conflict. In any case, the intent of myself and the scholars who have agreed to contribute to the writing of this book fits precisely within this climate of a partial "return of the repressed." Of course, we didn't seek in the thought of Russell, Jaspers, Aron, and Anders answers to our questions about the present or our fears, but rather examples of how one can think about and confront the problem of the nuclear threat with the insatiable restlessness of intelligence, and certainly without hiding the abyss it opens before us.

Two clarifications are necessary regarding the choices that have shaped our work. First: we are aware that the four figures at the center of this book are not the only ones who have addressed the issue of nuclear armaments with philosophical depth. The debate in Europe and America is vast and long-standing, and among its key figures are thinkers of the caliber of Elizabeth Anscombe, Norberto Bobbio, Dieter Henrich, and Lewis Mumford (to name just four examples). Thus, examining the ideas of Russell, Jaspers, Aron, and Anders is only a first step, an invitation to explore a topic that requires much broader and more in-depth investigations. The second clarification: we are well aware that the four figures in our book do not approach the issue of nuclear armaments in the same way. They do not, in fact, form a "school of thought," nor do they constitute a homogeneous intellectual community; rather, they represent very different approaches. But it is precisely for this reason that we chose them: we believe it would be fruitful to present many possible perspectives on the same problem and different ways of addressing it.

I conclude with a few words specifically about the profound differences that exist between the approaches of the four philosophers we are focusing on. First of all, Bertrand Russell (chapter 1): he was an intellectual who, like very few others in history, was able to advocate for peace, demonstrating the fallacy of the reasons that drive men to war. From 1914 until his death in 1970, he never stopped fighting against violence, which he viewed as both useless and the enemy of civilization and reason (in which he placed immense hope). Russell wrote short, incisive essays capable of convincing the common man, intellectuals, and political leaders alike. His stance in the face of the potential for nuclear holocaust was neither that of a politician (who calculates the real possibilities of action) nor that of a realist political scientist (who aims to describe the situation objectively); it was that of an engaged philosopher, a man who uses the best ideas available to explain how to pursue good (or avoid evil). In short, Russell's goal was to oppose the madness of his time with the awareness of a great intellectual and to develop a philosophy capable of shouldering the responsibility of such a challenge.

A very different attitude from that of Russell is taken by Aron (chapter 2), to whom we owe a monumental work on international relations, as is well known. In Aron's writings, a clear analysis of political dynamics prevails, along with a realism rooted in a concrete and dispassionate temperament. As Francesco Raschi highlights, Aron argued that even in the nuclear age, conflicts could and should remain limited, and did not necessarily lead to the global destruction feared by pessimists. The proof that this was possible, according to Aron, was provided by the Korean War, which, despite its destructive potential, remained confined to a limited geographical area and involved the use of conventional weapons. Certainly, Aron was aware of the possibility of a globally destructive atomic war and was deeply frightened by it; at the same time, however, he believed that if politics could keep conflicts within the bounds of reasonable and limited objectives (that is, not aimed at the destruction and unconditional surrender of the opponent), it would be possible to avoid the self-destruction of humanity.

Jaspers' perspective (chapter 3) is that of a philosopher, an heir to the great German tradition, not very inclined to attribute much importance to political dynamics, but attentive to the deeper reasons that underlie human behavior. Giunia Gatta explains that, for Jaspers, the solution to the problem of the nuclear threat was a spiritual and cultural revolution. It would not have been enough to find political or institutional solutions to the threat of atomic war; to eliminate it, something that was above politics would be needed, something that could radically change human behavior. In the concrete situation, this spiritual revolution had to tread the very narrow line that separated global destruction by atomic bombs on one side and the loss of freedom (that is, the condition for any choice) on the other. In other words, it was a matter of overcoming the difficult situation that emerged in 1945 without either falling into nuclear war or succumbing to the power of Soviet totalitarianism (for example, by renouncing the nuclear arsenal).

Günther Anders (chapter 4), too, who was a student of Husserl and Heidegger, was more inclined to reflect on philosophical issues than on power dynamics or the political-institutional paths to be taken. However, his affinity with Jaspers does not extend beyond this, as in *Die atomare Drohung*, published in 1983, he firmly rejects Jaspers' idea that the nuclear arsenal represents a useful tool for safeguarding freedom against totalitarianism. In Anders' view, the willingness to use atomic weapons was, in itself, already a concession to totalitarianism. Apart from his disagreement with Jaspers, the line of thought initiated by Anders, highlighted by Micaela Latini, concerns the gap between humans' enormous technical capabilities and their inability to envision and control the effects of the products they create. In this situation, human beings become marginal: unable to perform at the level of technological products, the individual becomes a mere cog in the service of a machine larger than themselves. In addition to presenting these ideas, Latini examines Anders' later reflections, when – after the Chernobyl disaster – he comes to theorize a form of counter-violence as an act of rebellion against those who promote and support the construction of products that destroy humanity.

Notes

1 N. Bobbio, "Prefazione alla seconda edizione," in *Il problema della guerra e le vie della pace* (Bologna, Il Mulino, 1984), 6.
2 Stockholm International Peace Research Institute, "States Invest in Nuclear Arsenals as Geopolitical Relations Deteriorate – New SIPRI Yearbook Out Now," June 12, 2023, available at www.sipri.org/media/press-release/2023/states-invest-nuclear-arsenals-geopolitical-relations-deteriorate-new-sipri-yearbook-out-now
3 Bobbio, *Il problema della guerra e le vie della pace*, 56.
4 Naturally, the episodes of terrorism that occurred in the 2000s are an exception; however, they do not, in themselves, constitute acts that can be categorized as warfare.

Bibliography

Bobbio, N., *Il problema della guerra e le vie della pace* (Bologna, Il Mulino, 1984).
www.sipri.org

1 Bertrand Russell's Commitment against Atomic Warfare

Alberto Castelli

Bertrand Russell and War

The thought on peace and the pacifist political commitment of Bertrand Russell date back to World War I. Beginning in 1914, he dedicated himself to highlighting the senselessness of war; he sought to unmask warmongering propaganda as an artificial rationalization of base instincts; he defended conscientious objectors in the name of freedom of thought; he stigmatized the readiness of writers, philosophers, and scientists to legitimize violence; and, in 1916, he wrote an open letter to American President Wilson, asking him to intervene politically to put an end to the hostilities.[1] Once the conflict ended, Russell did not abandon the ideas that had driven him to fight for peace; on the contrary, they remained at the core of his political thought until his death in 1970.[2]

This chapter does not aim to address Russell's pacifist journey as a whole, but only his reflections on wars fought with highly destructive weapons and on how to prevent them. His earliest articles on this topic date back to the 1930s, when it became clear to him that, due to the use of aerial bombing, war would no longer be a clash between armies but an indiscriminate massacre of civilians. "In a general war which ever side we take, England, France, Germany, and Italy, being densely populated and industrialized, will be quickly reduced to chaos by air raids and their consequences."[3] Since there was "no defence against attack from the air, except counter-attack," winning a war would have meant just killing "more women and children than they can kill of yours." Winners and losers, therefore, would share the same fate of chaos and destruction, without being able to gain any political or economic advantage.[4]

Thus, war was no longer just barbaric and violent, but also futile; for this reason, it was more important than ever that international disputes be settled not through armed conflict but through nonviolent legal means. To this end, it was necessary to give birth to a supranational authority, democratic and endowed with its own military force, that would enforce the law and strip states of the power to wage war. As Russell wrote in the famous essay *Which Way to Peace?* of 1936,[5] it was necessary to establish a "single supreme world government, possessed of irresistible force, and able to impose its will upon any national State or combination of States."[6]

DOI: 10.4324/9781003617174-2

Russell's thoughts on peace in the 1930s, therefore, were based on three beliefs: (1) war was senseless, not only because it was a breeding ground for barbarism but also due to the destructiveness of the military technology available to modern armies; (2) its deepest cause was the autonomy of states in foreign policy; and (3) it could be abolished through supranational institutions capable of limiting such autonomy. These beliefs must have appeared to Russell as confirmed by subsequent events: the spread of nationalism and World War II, the immense chaos caused by bombings, the atrocities, the destruction, and especially the explosion of the two atomic bombs that, on August 6 and 9, 1945, leveled Hiroshima and Nagasaki.

The Shock of the Atomic Bomb

The development of nuclear technology not only validated Russell's predictions and fears but also caused deep concern among scientists. Even before the bombings of Hiroshima and Nagasaki, the great Danish physicist Niels Bohr warned the British and American governments that the invention of the atomic bomb would initiate an extremely dangerous arms race among the victorious powers of the war. He therefore advised not to use it against Japan and to establish an international agreement for the control of nuclear energy. Bohr's opinion was shared by other academics as well; particularly, it was shared by a small group of scientists from the University of Chicago's Metallurgical Laboratory (the Met Lab), who were involved in the Manhattan Project. Under the guidance of Nobel Prize-winning chemist James Franck, between June 4 and 11, 1945, these scientists wrote a petition (later called the Franck Report) in which they asked the Secretary of War not to use the bomb against Japan, to merely demonstrate its power in an uninhabited area, and to establish an international agreement for the control of atomic energy.[7] Neither Bohr's opinions nor the Franck Report had any influence on the American government; the decision to use the bomb on Japan had already been made, believing that this would enable the United States to dictate its terms more easily at the end of the conflict.[8]

When Russell delivered his first relevant speech on the atomic bomb on November 28, 1945, at the House of Lords, he was aware both of the failure of scientists' attempts to influence the US government and of the need to insist on urging Western governments to seek "a way of cooperating with Russia."[9] Russell wrote:

> We must, I think, hope – and I do not think this is a chimerical hope – that the Russian Government can be made to see that utilization of these means of warfare would mean destruction to themselves as well as to everybody else. We must hope that they can be made to see that this is a universal human interest and not one on which countries are divided.[10]

The goal of this cooperation should have been, on one hand, the sharing of atomic technology information with the Russians (whose secret would have been short-lived anyway) and, on the other hand, placing nuclear technology under the control of the newly formed United Nations.[11] Once these goals were achieved, according to Russell, the problem of war should have been definitively resolved by laying the foundations for an International Authority that would compel states to resolve their disputes without resorting to force.

The Need for a Change

After the chaos caused by World War II, the idea of an International Authority that would prevent war gained significant prestige and led to both political mobilizations and a broad and in-depth body of literature on the subject. In Britain, for example, the Labour politician Henry Usborne organized World Government weeks in various cities, featuring mass rallies and prominent press coverage, where it was asserted that "the choice is between one world or none." Also, the president of the Board of Trade, Stafford Cripps, and the Foreign Secretary, Ernest Bevin, insisted that war had become collective suicide and advocated for the creation of a supranational authority to prevent the outbreak of conflicts. Similarly, Prime Minister Clement Attlee stated that if the world remained in its current state, it would eventually lead to a mutual annihilation.

In the United States, the pamphlet *One World* by the politician Wendell Lewis Willkie (who had previously run for US president in 1940) sold two million copies between 1943 and 1945, and the book *The Anatomy of Peace* by journalist Emery Reves became a bestseller, translated into 20 languages, between 1945 and 1950. The short essay by journalist and pacifist activist Norman Cousins, titled *Modern Man Is Obsolete*, also achieved notable success. One of the claims put forward by Cousins was that the need for a World Government existed even before August 1945, but that after the explosion of the atomic bombs, it had become an utter necessity. Another American intellectual, Robert Hutchins of the University of Chicago, even drafted a world constitution that circulated widely beyond academic circles.[12]

The political engagement of some groups of scientists must also be remembered. Among these, the Atomic Scientists' Association played an important role, and its Vice President József Rotblat would become significant in the pacifist efforts led by Russell in the following years.[13] In Britain, alongside and in competition with the Atomic Scientists' Association, was the Association of Scientific Workers, composed of scientists with decidedly more radical ideas and affiliated with the communist-led World Federation of Scientific Workers.[14] In the United States, in December 1945, the Federation of American Scientists was formed with the explicit goal of eliminating the threat of nuclear war. This association published a "Bulletin of the Atomic Scientists," to which Russell also contributed, and in 1946, the bestseller *One World or None*.[15]

Certainly, American political circles did not indulge in such "world-government dreams";[16] however, they did reflect on the need to establish control over nuclear technology. On June 14, 1946, the US representative to the United Nations Atomic Energy Commission (UNAEC), Bernard Baruch, proposed the creation of an International Atomic Development Authority, hoping to prevent the uncontrolled proliferation of nuclear weapons. Also in June 1946, US Secretary of State James Byrnes established an advisory committee that included, among others, the Under-Secretary of State Dean Acheson and the Chairman of the Tennessee Valley Authority, David Lilienthal. The task of this committee was to draft a report that the US government would be required to present to the UNAEC. The so-called Acheson-Lilienthal Report, written with the assistance of physicist Robert Oppenheimer, proposed the creation of an Atomic Development Authority that would hold a monopoly on fissile materials and distribute them only for peaceful purposes. Acheson and Lilienthal emphasized both that the atomic bomb posed an enormous danger, from which it was impossible to defend, and that it was not feasible to prevent the spread of the technology necessary to build it.[17]

As is well known, neither the proposals put forward by intellectuals nor those circulating in political circles were acted upon. As the opposition between the powers intensified, especially after the USSR tested its first nuclear bomb in October 1949, military superiority became the only objective to pursue.[18] Thus, when Niels Bohr, in his famous 1950 *Open Letter to the United Nations*, wrote that "a radical adjustment of international relationship is evidently indispensable if civilization shall survive," he was reflecting a deeply felt need, recognized at various levels but completely ignored by governments.

World Government and Western Values

A similar sentiment, of a rift between the awareness of the need for political and institutional renewal and the difficulty of achieving it, was present in the essays Russell wrote between 1950 and 1952. In *Obstacles to World Government*,[19] for example, Russell explained the situation in these terms:

> So long as there are different sovereign States, each with its own armed forces, each the unfettered judge of its own rights in any dispute, so long it is inevitable that there will continue to be wars from time to time.[20]

Until 1945, this situation in which "from time to time" a war would break out had been sustainable for humanity, albeit at the cost of high tolls in terms of deaths, destruction, and barbarism. However, once the weapons available to states had become so destructive as to jeopardize the survival of civilization, it was no longer possible to risk a large-scale conflict. Thus, it was extremely urgent to abolish the real cause of war, the autonomy of states in foreign policy, and to entrust the guarantee of peace to a supranational authority.

However, Russell identified two obstacles to this urgent reform. The first was related to the persistence of three major cleavages in the world: (1) the socio-economic cleavage, causing enormous disparities in resources between different regions of the planet; (2) what Russell termed "racial antagonism," concerning the difficulty of relations between very different peoples; and (3) the cultural cleavage, consisting of the ongoing presence of religious and ideological fanaticism. According to Russell, such cleavages could have been significantly reduced if, for a sufficiently long period, peoples had enjoyed "security, prosperity, and liberal education."[21] In other words, if it had been possible to ease political tensions, calm tempers, and demonstrate the advantages of a system without aggressive sovereign states, the establishment of an International Authority would no longer have seemed so unachievable.

The second obstacle to overcoming the state was related to the sentiment of national belonging. In the article "The Kind of Fear We sorely need," published on October 29, 1950, in the *New York Times* (*Magazine*),[22] Russell explained that the sentiment of national belonging was deeply rooted in the human spirit because it was connected to the "love of what is familiar" (for example, food, language, and acquired habits), and to the "desire to dominate and the fear of being dominated."[23] However, even this deeply binding sentiment could have been overcome if two simple truths had been made clear: (1) that, under certain conditions, it tends to transform into aggressive and bellicose nationalism; and (2) that, since war is the greatest and deadliest enemy of civilization, it is also an enemy of everything that the national sentiment seeks to preserve. According to Russell, once peoples understood and accepted these two truths, the fear of war and the desire to preserve their civilization would drive them to support a World Government, whose need was urgent.

To argue that peoples need to fear war, however, did not mean, for Russell, that peace was an absolute value to which every other value should be sacrificed. In particular, it would have been wrong to achieve peace at the price of "subjugation" to Soviet totalitarianism and the renunciation of Western civilization and values. Russell dealt with this topic in *Western Values*,[24] published in 1952:

> We of the Western nations, have at the present time a very tremendous responsibility. We have discovered the way of life which eliminates many ancient evils that used to be thought part of the inevitable lot of man. We know how to produce communities without serious poverty, without plagues, pestilences, or famines, with a high general level of education and a very small proportion of crime.[25]

All this could not be sacrificed; rather, it had to be defended with arms if necessary. Even an atomic threat appeared acceptable to Russell if it were a useful deterrent; that is, if it dissuaded the Soviets from unleashing a new war, bought time, eased tensions, and initiated negotiations.[26]

To these considerations, however, Russell added that Soviet "scientific totalitarianism" was not the only enemy of Western values; the willingness of Westerners themselves to fully succumb to the fanatic logic that wars bring with them could also trigger a dangerous process of regression. Russell wrote:

> Those who wish to combat what is evil in the Soviet régime must energetically combat every germ of similar evil among ourselves. There would be little purpose in a long and devastating war which ended by the victors adopting all the vices of the vanquished.

Thus,

> if [...] an ideological war is forced upon us, we must remember, throughout, that a real victory must be ideological as well as military. We have something of infinite value to fight for. Let us make sure that we do not lose it in the heat of battle.[27]

In conclusion, it can be argued that for Russell, the overlapping needs at the beginning of the 1950s were fourfold: (1) to ease international political tensions; (2) to initiate a process of dialogue and cooperation to establish an International Authority; (3) to reduce economic disparities and convince every nation that the real enemy is war, not neighboring peoples; and (4) to defend Western values both against Soviet totalitarianism and against the willingness of Westerners themselves to sacrifice these values on the altar of war.

The H-Bomb and Its Enemies

The Bikini test on March 1, 1954, in which the United States tested a hydrogen bomb with a yield of 15 megatons, placed the nuclear threat on a new level. A single device of this type had a power approximately five hundred times greater than the bomb dropped on Hiroshima and was capable of completely destroying large cities like London or Moscow. The detonation at Bikini was heard over 300 kilometers away, and radioactive dust spread over a vast area, causing casualties and serious illnesses among American military personnel, the inhabitants of the Marshall Islands, and the crew of the Lucky Dragon No. 5, a Japanese fishing vessel. In short, the Bikini test was the definitive proof that atomic technology could cause damage to the entire planet and have lethal consequences for humanity. Moreover, along with the hydrogen bombs, fast missiles with great range were invented, which could directly strike the United States, no longer protected by its geographical isolation.[28]

In this new situation, in March 1954, the National Security Council of the United States tasked James R. Killian, president of the Massachusetts Institute of Technology, with conducting a study on the United States' exposure to a potential nuclear "first strike." In the report, which was delivered in February 1955, Killian concluded that the United States was highly vulnerable. He, therefore, suggested investing substantial sums in the construction of intercontinental ballistic missiles to achieve an improvement, albeit relative and temporary, in defensive capability.[29] Meanwhile, in June 1954, the British government also decided to build the H-bomb.

In general, the fear of Soviet nuclear power and strong government pressure pushed public opinion in the United States and Britain to support investments in atomic weapons. However, there were also sharply critical voices (the British Council of Churches, the Trade Unions) and associations that continued to promote the establishment of a World Government for the control of atomic weapons and for the abolition of war (among these, in Britain, the Parliamentary Group for World Government was certainly the most important).[30]

On one hand, the Bikini test made evident the enormous threat to which humanity was exposed; on the other, it spurred a series of efforts by politicians and scientists to mobilize public awareness. These efforts, though remaining marginal, demonstrated a deep understanding of the gravity of the situation, outlined the necessary measures to avoid nuclear holocaust, and emphasized the urgency of exerting direct pressure on the world's powerful leaders. In this context, Russell's reflections on the H-bomb took shape.[31]

"Remember Your Humanity and Forget the Rest"

Russell's most evocative essay of this period is, without a doubt, *Man's Peril*. The text of the essay was read in a BBC broadcast on December 23, 1954, and printed in the BBC weekly *The Listener* under the title "Man's Peril from the Hydrogen Bomb" on December 30, 1954.[32] Russell immediately made it clear that, in the face of the threat of a nuclear holocaust, it was necessary to overcome all political divisions and take responsibility for the fate of humanity as a whole. To address the problem of the atomic bomb, it was necessary to think not as a citizen of a state or as a member of a political party, but only "as a human being, a member of the species Man, whose continued existence is in peril." On the contrary, the leaders of the two superpowers behaved like duelists, aware of the danger they faced and the senselessness of their struggle, but unable to find a way out for fear of appearing weak and cowardly. The only hope was that the duelists would develop a new consciousness and take responsibility for the situation. At the end of the article, Russell wrote:

We appeal, as human being to human beings: Remember your humanity, and forget the rest. If you can do so, the way lies open to a new paradise; if you cannot, there lies before you the risk of universal death.[33]

The following year, in the essay entitled "The Road to Peace," Russell thought about how to urge the duelists to take more responsible positions.[34] He argued that it would be appropriate to publish a "statement by a small number of men of the highest scientific eminence as to the effect to be expected from a thermonuclear war." It should have been a technical report, avoiding any moral or political judgment, and clarifying the destructive power of the bombs and the effects of radioactivity, even on future generations.

Such a statement should have finally made it clear that "a thermonuclear war would not bring victory to either side, and would not create the sort of world desired by Communists or the sort of world desired by their opponents or the sort of world that uncommitted nations [desired]."[35] It should have clarified, in short, that atomic bombs had no utility except as deterrents, that "they [were] useful only if not used," and that it made no sense to maintain an expensive and lethally dangerous apparatus whose purpose was not to be used. Explaining all this should have urged politicians and public opinion towards a more reasonable attitude, open to détente, and even to a "temporary armistice" during which to try to find agreements on a disarmament program.[36]

The ultimate goal of the peace process, of course, should have been the establishment of a World Authority, on whose structure and organization, in "The Road to Peace," Russell dwelled, providing a few more details than he had in the past. He envisioned a World Authority composed of subordinate Federations (approximately homogeneous in size and population), which in turn would be composed of states. The subordinate Federations, not the states, should have appointed the delegates to the representative bodies of the World Authority. Russell hypothesized that it would be appropriate to form eight Federations: the United States, the Soviet Union and its European satellites, the British Commonwealth, China with some satellites of the communist Far East, Latin America, Latin Europe (Italy, France, Spain, Portugal, and possibly Belgium), the Muslim world, and Germany with Scandinavia, Switzerland, the Netherlands, and Austria.

The powers of the World Authority were supposed to concern international treaties and any issues related to the prevention and resolution of violent conflicts. Thus, the powers of the World Authority were supposed to extend to all problems that, if poorly managed, would be sources of conflicts and wars. Among these, according to Russell, was the issue of the disparity of resources and standards of living between different regions of the planet. To the World Authority, therefore, should have fallen the task of promoting a fairer distribution of resources among all countries, preventing "economic envy" from being "a perpetual incitement to hostility," and the West from being "accused, whether justly or not, of economic imperialism."[37]

To pursue such ambitious goals, the World Authority would have needed to possess a military force that was effective and superior to that of any other state or subordinate Federation. Russell wrote that

> the national forces of separate States should be reduced to the dimensions necessary for the preservation of internal order. They must not be allowed to have any of the more effective weapons of war. The more effective weapons must be a monopoly of the international armed forces. I do not think that even the international armed forces ought to possess such a weapon as the hydrogen bomb, because this is capable of bringing death to others than those against whom it is directed. But I do think that the international force ought to possess whatever weapons are necessary to make it certain victory against rebels.[38]

According to Russell, the project of establishing such a World Authority could be subject to two criticisms. The first was that it would limit freedom of states (and consequently of peoples); the second concerned the danger that the concentration of military power would give rise to a global tyranny. Regarding the first criticism, Russell admitted that the World Authority would restrict freedom of states, but he added that, on one hand, this limitation would be balanced by the benefits of a stable and lasting peace; and that, on the other hand, it would have been an acceptable limitation because "the World Authority should leave each national State and each subordinate Federation complete freedom in everything not affecting peace."[39] Regarding the second criticism, Russell pointed out that, in developed countries, it was rare for political power to be subjugated by military power and that "the same methods by which civilian control over national armed forces has been established may be expected to be equally effective against an international armed force."[40]

The Intellectuals' Engagement

Since 1955, Russell undertook a series of initiatives to persuade the leaders of the nuclear powers to initiate a policy of détente. First of all, he drafted the text that would become known as the *Russell-Einstein Manifesto*, signed not only by Russell and Einstein but also by nine other Nobel Prize winners.[41] The *Manifesto* was sent to US president Dwight Eisenhower, Soviet premier Nikolai Aleksandrovich Bulganin, Chinese premier Zhou Enlai, French president René-Jules-Patrice-Gustave Coty, Canadian prime minister Louis St. Laurent, and British prime minister Anthony Eden (who was the only one to send Russell a formal letter of appreciation). Essentially, the *Manifesto* was a synthesis of *Man's Peril*, published just a few months earlier: it emphasized that the issue of nuclear energy concerned every human being and called for it to be addressed setting aside all political divisions.[42] As is well known, the appeal of Russell and Einstein went entirely unheard by the political leaders, and indeed, the arms race saw a significant escalation: in 1956, the United States conducted 18 nuclear tests and the Soviets nine; in 1957, the United States conducted 32 and the Soviets 15; and in 1958, the United States 77 and the Soviet Union 29. Moreover, Britain successfully developed the H-bomb, testing it on May 15 and 31, and June 19, 1957, raising fears of a rapid and uncontrollable proliferation of nuclear weapons worldwide.[43]

Faced with all this, Russell, along with physicists Józef Rotblat and Cecil Powell, organized the first Pugwash Conference on Science and World Affairs, with the stated goal of bringing together renowned scientists to discuss both technical issues and how to initiate a peace process.[44] The Pugwash Conference began on July 7, 1957, in the Canadian town. Russell did not attend in person due to health reasons. However, at Rotblat's suggestion, he sent a greeting message, which was later published in the *New York Times* and the *Montreal Gazette* on July 10.[45] The message opened with the observation that, two years after the *Russell-Einstein Manifesto*, the risk of a nuclear disaster had actually increased. For this reason, in Russell's intentions, the meeting of eminent scientists – from various countries and invited to Pugwash solely for their scientific expertise – was meant to demonstrate that it could be "the seed from which, gradually, a sense of common human problems" would replace "the present futile competition, from which nothing but a catastrophe can result."[46]

Twenty-two scientists took part in the Pugwash Conference: seven Americans, three Soviets, three Japanese, two Britons, two Canadians, and one each from Australia, Austria, China, France, and Poland. Among them were physicists, biophysicists, chemists, a geneticist, a physiologist, a psychiatrist, and a jurist. The number of Western scientists was significantly higher than that of scientists from communist countries, but Russell still considered the presence of the latter to be significant. He was, however, disappointed by the absence of Indian scientists, who, as representatives of a major neutral power, could have played an important mediating role.

During the Pugwash Conference, discussions covered the risks associated with radiation, the possibility of controlling atomic weapons and initiating disarmament, and the social responsibilities of scientists. The most widely agreed-upon results concerned the effects of radiation on humans, as these were objectively observable phenomena. In contrast, for the other topics discussed, only general reports were produced. The final text of the Conference demonstrated a strong continuity with the Russell-Einstein Manifesto and emphasized that full agreement had been reached on the fundamental aims to pursue for the common well-being of humanity.[47]

Russell also sought to directly influence political leaders by writing an open letter to Eisenhower and Khrushchev, reminding them that their common interests were more important and numerous than what divided them. He emphasized that cooperation to prevent nuclear war was both possible and necessary.[48] At the same time, Russell also launched the Campaign for Nuclear Disarmament, which was presented at a meeting at the Central Hall in Westminster on February 17, 1958. Through this Campaign, he hoped to reach a broader audience than that reached by his articles and essays. The Campaign for Nuclear Disarmament advocated for unilateral British disarmament, based on the belief that "if Great Britain gave up her part in the nuclear race and even demanded the departure of United States bases from her soil, other nations might follow suit."[49]

Overcoming Fanaticism and Nationalism

By the late 1950s, Russell's ideas on peace and the nuclear threat found a point of synthesis in the essay *Common Sense and Nuclear Warfare*. Russell stated that the presence of atomic weapons itself constituted a problem and a source of conflict. Indeed, they forced states into an expensive and never-ending arms race, sustained through heavy taxation, strict social control measures, and, inevitably, a constant campaign of hatred against the enemy. The mere presence of atomic weapons, therefore, brought with it a brake on economic growth, a systematic restriction of freedom, and a climate of fear and collective hysteria that would ultimately make conflict inevitable.

To escape this dangerous and senseless situation, Russell reiterated that the illusion that atomic weapons could be used rationally must finally be abandoned. He noted that, according to military estimates, even in a "victorious" nuclear confrontation, H-bomb explosions on US territory would result in 72 million deaths and 21 million injuries (from a population of 150 million people in 1950). Russell also observed that these figures had to be augmented by the victims of fallout (the deposition of radioactive ashes), the deaths caused by the lack of medical care (since the healthcare system would be destroyed), the absence of clean water and healthy food, the spread of epidemics, and the collapse of social cohesion.

The world which would emerge from a nuclear war would not be such as is desired by either Moscow or Washington. On the most favorable hypothesis, it would consist of destitute populations, maddened by hunger, debilitated by disease, deprived of the support of modern industry and means of transport, incapable of supporting educational institutions, and rapidly sinking to the level of ignorant savages.[50]

Russell further pointed out that these effects would not be confined to the warring states but would impact the entire planet.

For Russell, the realization that war was futile had to represent the first step towards a solemn joint declaration by the United States and the USSR to the effect that they will seek to settle their differences otherwise than by war or the threat of war, and that, to implement this declaration, they should appoint a permanent joint body to seek measures tending towards peace and not altering the existing balance of power.[51]

Such a "permanent joint body," in turn, would have to be an intermediate step towards the International Authority, which Russell had discussed on many occasions and whose characteristics have already been examined previously.

In the tenth chapter of *Common Sense and Nuclear Warfare*, Russell revisited the factors that impeded détente and the building of peace. First and foremost, he mentioned fanaticism, which drove the belief that the extermination of human life on the planet was preferable to the victory of the opposing side. Russell wrote: fanatics

> maintain that the evils inflicted by the Kremlin or by Wall Street, as the case may be, are so great that, in a world dominated by either, life would not be worth living and it would be a kindness to future generations to prevent them from being born.[52]

Against this argument, Russell asserted that no political regime, no matter how tyrannical, would last forever; that in every system and in every era, good and evil were intertwined; and that it was unreasonable to believe that nothing good could ever occur within a given system. Thus, "only a man who [...] lost his sense of human values" could truly believe that it was better to die than to live under an undesirable political regime.[53]

The second obstacle to peace mentioned by Russell was nationalism, understood not as pride in one's origins and traditions, but as "collective self-glorification" and "conviction that it is right to pursue the interests of one's own nation however they may conflict with those of others."[54] Russell explained very effectively the reasons why, in his view, this nationalism was absurd:

> What should we think of an individual who proclaimed: "I am morally and intellectually superior to all other individuals, and, because of this superiority, I have a right to ignore all interests except my own"? There are, no doubt, plenty of people who *feel* this way, but if they proclaim their feeling too openly, and act upon it too blatantly, they are thought ill of. When, however, a number of such individuals, constituting the population of some area, collectively make such a declaration about themselves, they are thought noble and splendid and spirited. They put up statues to each other and teach schoolchildren to admire the most blatant advocates of the national conceit.[55]

Russell commented that until 1945, humanity had lived with "collective self-glorification," paying the price in terms of wars and suffering, but it had not been destroyed by it. In the nuclear age, however, indulgence towards such nationalism risked bringing the human race to extinction and, therefore, it had to be finally abandoned.

Along with nationalism, nationalist education, which instilled the desire to fight for the greatness of the state, also needed to be abandoned:

> It is no longer to the interest of any country to emphasize its superiority to other countries or to cause its boys and girls to believe it invincible in war. Nor it is a good thing to present martial glory as what is, above all things, to be admired.[56]

Therefore, the educational system needed to be profoundly reformed, starting with the way history was taught (with all traces of chauvinism to be removed). The task of education should have become "to make vivid in the minds of the young both the merits of a civilized way of life and the needless dangers to which it is exposed by the survival of competitive ideals which have become archaic."[57] In short, it was a matter of replacing fear with hope and the desire to prevail with the desire to cooperate.

The Future of Man

The last essay worth considering is *Has Man a Future?* from 1961, in which Russell not only revisited the themes developed over the previous decades but also devoted insightful pages to arguing that peace and the progress of civilization were achievable goals.[58] Making this point clear meant, for Russell, asserting that political choices were not exclusively determined by madness or the will to power, and that reason and foresight could also contribute to shaping reality (and that, since humans possessed a bomb "a thousand times more powerful than the A-Bomb," it had become necessary for them to be followed).[59]

Reason and foresight, according to Russell, advised pursuing four objectives: (1) nuclear disarmament, (2) the abandonment of nuclear tests, (3) the recognition that the balance of terror was highly risky, and (4) the non-proliferation of nuclear weapons among powers that did not possess them, so as not to multiply the chances of an atomic war. Russell explained that the advisability of these four objectives was "universally admitted," but they were not being realized because, instead of a "sane" attitude, a "mutual enmity" continued to prevail, rooted in pride, suspicion, fear, and the love of power.[60]

These sentiments, according to Russell, gave rise to a sort of omnipotence syndrome that tended to take hold of politicians and to lead nations to disaster. It was thus the omnipotence syndrome that drove Khrushchev to claim he wanted to destroy the West, and Dulles to assert that nuclear war could be won by the

Americans. It drove them towards "an utter folly, even from the narrowest point of view of self-interest," because "ruin, misery, and death throughout one's country as well as that of the enemy [was] the act of a madman."[61]

Once again, however, for Russell, this "utter folly" was not an insurmountable fate; violent behaviors did not appear to him to be inscribed in the unchangeable depths of human beings. On the contrary, "what [was] called 'human nature'" was "in the main, the result of custom and tradition and education, and, in civilized men, only a very thin fraction [was] due to primitive instinct." Thus, if a way could be found to prevent future generations from experiencing violence and learning the logic of "mutual enmity," war could become as absurd as dueling, slavery, and other now-obsolete practices. "No doubt there would still be some homicidal maniacs, but they would no longer be heads of Governments."[62]

Also in *Has Man a Future?* Russell insisted that the only way to abolish war and ensure the survival of the human race in the nuclear age was to create an International Authority. This authority would need to grant the greatest freedom to federated states in managing their internal affairs, but it should also have the capacity to punish any violation of the law by any national state. Compared to previous discussions on the topic, however, in *Has Man a Future?* Russell took care to clarify the conditions that would ensure the effectiveness of the World Authority. Firstly, a "vigorous inculcation of loyalty" was necessary, not based on the fear of an external threat, but on the commitment to address the dangers still affecting humanity, such as poverty, hunger, disease, and, above all, war and nuclear holocaust. Russell was aware that fear of external threats was the best instrument of "social cohesion"; however, he argued that it would be "unduly pessimistic to suppose that nothing more positive and more beneficial could take its place."[63] The support for the International Authority would also need to be based on the advantage that a growing industrial system required ever-larger markets and, therefore, would benefit greatly from a unified and peaceful world. It should also be added that the increased speed of communications and transportation, along with the rising cost of weapons, would contribute to pushing towards larger political and administrative units.

In short, for Russell, building peace was difficult but not impossible, and nuclear war was not an inevitable fate, but a choice. One could choose to surrender to the power of weapons or to control them so that they did not destroy civilization and the human race.

> We must become aware that the hatred, the expenditure of time and money and intellectual ability upon weapons of destruction, the fear of what we may do to each other, and the imminent [...] risk of an end to all that man has achieved – we must be aware, I say, that all this is a product of human folly. It is not a decree of Fate. It is not something imposed by natural conditions. It is an evil springing from human minds, rooted in ancient cruelty and superstition.[64]

The Strength of Reason and the Reasons of Strength

The political and intellectual commitment that Russell was able to put into practice to avert the danger of nuclear war and establish lasting peace was not only intense and sustained over time but also characterized by a rare communicative effectiveness, consistent with much of the international mobilization of scientists and intellectuals, and rooted in a solid and long-standing philosophical-political tradition – the British one.

The relationship between Russell's ideas and the British philosophical-political tradition deserves an in-depth analysis, which cannot be undertaken here. Let it suffice to mention that Russell's ideas took shape within the context of a broad and profound debate on the economic, political, and cultural causes of war and on how to build peace. To convey the importance and breadth of the debate, let it be permitted to recall that, between the beginning of the 20th century and the outbreak of World War II, leading intellectuals such as John A. Hobson, John Maynard Keynes, Goldsworthy Lowes Dickinson, Leonard Trelawny Hobhouse, Harold Laski, Norman Angell, Leonard Woolf, Barbara Wootton, and Lionel Robbins took part in it.[65]

Russell, as noted earlier, was part of this debate, was familiar with the works of many of the authors mentioned above, and was therefore aware of the complexity of international relations. However, he chose to write short, incisive essays capable of convincing both the man in the street and intellectuals and political leaders. He believed he could contribute to the building of peace by, on one hand, spreading reasonable ideas and accurate data; and on the other, exposing the baselessness of passions and the fallacy of arguments that drive people to fight. That is why he devoted so much attention to the psychological and cultural factors of war (such as fanaticism, superstition, pride, mutual suspicion, and the syndrome of omnipotence), and to refuting the old belief that war could be politically or economically advantageous (as Norman Angell had also tried to do before him).

One might perhaps accuse Russell of excessive faith in reason and an optimism that fears confronting the depths from which violence among humans draws its strength. However, it must be clear that his position regarding the possibility of a nuclear holocaust is not that of a politician or a realist political scientist (interested in describing the objective situation or calculating immediate concrete possibilities); it is that of a committed philosopher, of a person who applies the best ideas available to explain how one could – if one had the will – pursue the good (or avoid the evil).

Russell attempted to counter the madness driving (and still driving) all of humanity towards self-destruction with his awareness as a great intellectual. In a difficult and terrifying situation, facing the first of what we now call global issues, Russell sought to develop a thought capable of taking responsibility for it. He took on the task that – if personal opinion is permitted – should be that of every scholar: to use intelligence and knowledge to let the strength of reason prevail over the reasons of strength.

Notes

1. B. Russell, "An Appeal to the Intellectuals of Europe (1915)," in *Justice in Wartime* (Chicago – London, The Open Court Publishing Co., 1916), 1–19.
2. As is known, Russell's pacifist commitment meant for him the ruin of his public reputation, a setback in his career, imprisonment, and even, at times, a threat to his physical safety. See R.A. Rempel, "Introduction," in *The Collected Papers of Bertrand Russell*, eds. Bernd Frohmann, Mark Lippincott, Richard A. Rempel (London and New York, Routledge, 1988, vol. 13), xiv–xvi. For the philosophical journey that led Russell towards pacifism, including his readings of William James and Norman Angell, see P. Ironside, *The Social and Political Thought of Bertrand Russell* (Cambridge, Cambridge University Press, 1996), 85–88.
3. B. Russell, "How to Keep the Peace (1935)," in *The Collected Papers of Bertrand Russell*, ed. E. Bone and M.D. Stevenson (London – New York, Routledge, 2008, vol. 21), 35.
4. B. Russell, "The Prospect of Permanent Peace, (1935)," in *The Collected Papers of Bertrand Russell*, ed. Andrew G. Bone vol. 21, 82. This essay is derived from a lecture given on November 28, 1935 - during the height of the crisis of the Italian invasion of Abyssinia - at the Fabian Society at Friends Hall in London. See also: B. Russell, *Britain Must Be Neutral* (1935), now under the title *How Not to Fight Fascism* in *The Collected Papers of Bertrand Russell*, vol. 21, 37–39; and the article written by Russell for *Peace News*, which was published on May 29, the day after the bombing of Guernica, *Humanizing Warfare* (1937), now in *The Collected Papers of Bertrand Russell*, vol. 21, 508–510. The thesis that it is impossible to achieve useful goals through modern warfare is widespread in British political thought and dates back to before the use of aerial bombardments. Its most famous formulation is probably that of Norman Angell in *The Great Illusion* of 1910. Regarding Angell, allow me to refer you to: A. Castelli, *The Peace Discourse in Europe 1900–1945* (London – New York: Routledge, 2019), 21–37.
5. B. Russell, *Which Way to Peace?* (London, Michael Joseph, 1936). About *Which Way to Peace?* See: A. Ryan, *Bertrand Russell: A Political Life* (London, Penguin Books, 1988), 145–155.
6. Russell, *Which Way to Peace?* 173.
7. It should also be remembered that The Franck Report followed a memorandum that had been unsuccessfully presented to War Secretary of State James Byrnes by the physicist Leo Szilard, associate at the Met Lab and co-builder, with Enrico Fermi, of the first nuclear reactor. Russell talked about the Franck Report as a "very admirable and far-seeing document" in *Has Man a Future?* (London, Penguin Books, 1961), 17. About the Franck Report, see M. Price, "Roots of Dissent: The Chicago Met Lab and the Origins of the Franck Report," *ISIS: The History of Science Society* 2 (June 1995), 224–244. See also L.S. Wittner, *Confronting the Bomb: A short History of the World Nuclear Disarmament Movement* (Stanford, Stanford University Press, 2009), 3–6.
8. Wittner, *Confronting the Bomb*, 5–7; P. Boyer, *By the Bomb's Early Light. American Thought and Culture at the Dawn of the Atomic Age* (Chapel Hill, The University of North Carolina Press, 1994), 49–58.
9. The speech is reproduced in full in Russell, *Has Man a Future?* 19–24.
10. Russell, *Has Man a Future?* 22.
11. Russell, *Has Man a Future?* 24. See also B. Russell, *The Autobiography of Bertrand Russell* (London, Allen & Unwin, 1969, vol. III), 16–18.
12. Wittner, *Confronting the Bomb*, 10, 14 and 15–16. About the American debate, see: P. Boyer, *By the Bomb's Early Light*, 33–45.
13. For more on Rotblat, see the essays published in Reiner Braun, Robert Hinde, David Krieger, Harold Kroto, and Sally Milne, *Joseph Rotblat: Visionary for Peace* (Weinheim, Wiley VCH Verlag GmbH & Co., 2007).

14 A.G. Bone, "Introduction," in *The Collected Papers of Bertrand Russell*, ed. A.G. Bone (London – New York: Routledge, 2020, vol. 28), xli. See also: J. Wang, *American Science in an Age of Anxiety: Scientists, Anticommunism, and the Cold War* (Chapel Hill, University of North Carolina Press, 1999), 253–288.
15 See Boyer, *By the Bomb's Early Light*, 76–81.
16 Boyer, *By the Bomb's Early Light*, 33.
17 J.P. Baratta, "Was the Baruch Plan a Proposal of World Government?" *The International History Review* 7, no. 4 (1985), 592–621.
18 Wittner, *Confronting the Bomb*, 30–32 and 37–38.
19 The article is a summary of a series of lectures he gave at the Assembly Hall in Sydney, in coordination with the Australian Institute of International Affairs (AIIA), between June 26 and July 5, 1950. The text is now published in *The Collected Papers of Bertrand Russell*, ed. A.G. Bone (London – New York, Routledge, 2020, vol. 26), 46–61.
20 B. Russell, "Obstacles to World Government," in *The Collected Papers of Bertrand Russell*, ed. A.G. Bone (London – New York, Routledge, 2020, vol. 26), 46.
21 B. Russell, *Obstacles to World Government*, 61.
22 B. Russell, "The Kind of Fear We Sorely Need," in *On Nationalism in The Collected Papers of Bertrand Russell*, ed. Andrew G. Bone (1950, vol. 26), 359–366.
23 Russell, *On Nationalism*, 363.
24 Russell, *Western Values*, a text read in a BBC Pacific Service radio broadcast titled *Calling Australia*, recorded on March 10, 1952, now in *The Collected Papers of Bertrand Russell*, vol. 26, 438–442.
25 Russell, *Western Values*, 441.
26 See: B. Russell, *How Near Is War?* (London, Derrick Ridgway 1952), now in *The Collected Papers of Bertrand Russell*, vol. 26, 443–461.
27 Russell, *Western Values*, 442.
28 See: D. Holloway, *Stalin and the Bomb. The Soviet Union and Atomic Energy 1939–1956* (New Haven, Yale University Press, 1995), 306–308. See also: A.G. Bone, "Introduction," xvii–xviii.
29 The Killian Report can be read on the US Department of State's website at the following address: https://history.state.gov/historicaldocuments/frus1955-57v19/d9.
30 See: A.G. Bone, "Introduction," xxxv; and Wittner, *Confronting the Bomb*, 58–59. For further reading: L. Wittner, *The Struggle against the Bomb* (Stanford, Stanford University Press, 1993), 269–293; H. Nehring, *Politics of Security: British and West German Protest Movements and the Early Cold War, 1945–1970* (Oxford, Oxford University Press, 2013), 41–61.
31 See, for example: B. Russell, "The Morality of 'Hydrogen' Politics," in *The Collected Papers of Bertrand Russell*, ed. Andrew G. Bone (1954, vol. 28), 51; and B. Russell, "The Hydrogen Bomb," in *The Collected Papers of Bertrand Russell*, ed. Andrew G. Bone (1954, vol. 28), vol. 28, 22.
32 B. Russell, "Man's Peril," in *The Collected Papers of Bertrand Russell*, ed. Andrew G. Bone (1954, vol. 28), 82–89.
33 B. Russell, *Man's Peril*, 89.
34 B. Russell, "The Road to Peace," in *The Bomb: Challenge and Answers*, ed. G. McAllister (London – Batsford, 1955), 47–68; now in *The Collected Papers of Bertrand Russell*, vol. 28, 352–355. Gilbert McAllister was a Labor politician and secretary-general of the World Association of Parliamentarians for World Government.
35 B. Russell, "The Road to Peace," 357.
36 B. Russell, "The Road to Peace," 365.
37 B. Russell, "The Road to Peace," 371.
38 B. Russell, "The Road to Peace," 369.

39 B. Russell, "The Road to Peace," 367.
40 B. Russell, "The Road to Peace," 369.
41 For information on Einstein's participation in this project, see *The Autobiography of Bertrand Russell*, vol. III, 74–78. See also the letter by which Russell invites Einstein to participate in the project in *The Selected Letters of Bertrand Russell. The Public Years, 1914–1970*, ed. N. Griffin (London – New York, Routledge, 2001), 488–490. The *Manifesto* was presented at a press conference organized by the "Observer," held on July 9, 1955. The other signatories of the *Manifesto* were: Max Born, P.W. Bridgman, L. Infeld, Fréderic Joliot-Curie, Hermann J. Muller, Linus Pauling, C.F. Powell, J. Rotblat, and Hideki Yukawa.
42 The *Manifesto* is published in B. Russell, *Has Man a Future?* 55–58.
43 See A.G. Bone, "Introduction," in *The Collected Papers of Bertrand Russell*, ed. B. Russell, vol. 29, xv–xvi. On the proliferation of nuclear tests, in March 1957, Russell wrote an article titled "Should H-bomb Tests Be Continued?" in which he strongly advocated for the suspension of nuclear tests. B. Russell, "Should H-bomb Tests Be Continued?" *The New Scientist*, March 28, 1957, now in *The Collected Papers of Bertrand Russell*, vol. 29, 310–315.
44 A Rotblat e alla Pugwash Conference sarà attribuito il Premio Nobel per la pace nel 1995. Sulle origini della Pugwash Conference, con particolare riferimento al ruolo svolto da Joliot-Curie, see G. Roberts, *Science, Peace and Internationalism: Frédéric Joliot-Curie, the World Federation of Scientific Workers and the Origins of the Pugwash Movement*, in Alison Kraft and Carola Sachse, edited by, *Science, (Anti-)Communism and Diplomacy. The Pugwash Conferences on Science and World Affairs in the Early Cold War* (Leiden – Boston, Brill, 2020), 43–79.
45 B. Russell, *Message to First Pugwash Conference* (1957), now in *The Collected Papers of Bertrand Russell*, vol. 29, 341–345.
46 B. Russell, *Message to First Pugwash Conference* (1957), now in *The Collected Papers of Bertrand Russell*, vol. 29, 344.
47 About the Pugwash Conference, see A.G. Bone, *Introduction*, in B. Russell, in *The Collected Papers of Bertrand Russell*, vol. 29, xlii–lii. See also J. Rotblat, *Bertrand Russell and the Pugwash Movement. Personal Reminiscences*, in http://digitalcommons.mcmaster.ca/viewcontent.cgi/context=russelljournal
48 Answers by Khrushchev and John Foster Dulles will be published in *The Vital Letters of Russell, Khrushchev, Dulles*, introduction by Kingsley Martin (London, MacGibbon & Kee, 1958). See: B. Russell, *The Autobiography of Bertrand Russell*, vol. III, 102.
49 B. Russell, *The Autobiography of Bertrand Russell*, vol. III, 104. See Wittner, *Confronting the Bomb*, 58–60.
50 B. Russell, *Common Sense and Nuclear Warfare* (London, Allen & Unwin, 1959, now also London – New York, Routledge, 2001), 31.
51 B. Russell, *Common Sense and Nuclear Warfare*, 2001, 27.
52 B. Russell, *Common Sense and Nuclear Warfare*, 61.
53 B. Russell, *Common Sense and Nuclear Warfare*, 63.
54 B. Russell, *Common Sense and Nuclear Warfare*, 64.
55 B. Russell, *Common Sense and Nuclear Warfare*, 65.
56 B. Russell, *Common Sense and Nuclear Warfare*, 67.
57 B. Russell, *Common Sense and Nuclear Warfare*, 70.
58 B. Russell, *Has Man a Future?* 14.
59 B. Russell, *Has Man a Future?* 30–31.
60 B. Russell, *Has Man a Future?* 31.
61 B. Russell, *Has Man a Future?* 45.
62 B. Russell, *Has Man a Future?* 47.
63 B. Russell, *Has Man a Future?* 83.

64 B. Russell, *Has Man a Future?* 109.
65 Castelli, *The Peace Discourse in Europe 1900–1945*; see also Alberto Castelli, *Una pace da costruire* (Milan, FrancoAngeli, 2002), 25–81.

Bibliography

Angell, Norman. *The Great Illusion* (London, William Heinemann, 1910).
Baratta, Joseph Preston. "Was the Baruch Plan a Proposal of World Government?" *The International History Review* 7, no. 4 (1985), 592–621.
Bone, Andrew G. "Introduction," in *The Collected Papers of Bertrand Russell*, ed. A.G. Bone (London – New York, Routledge, 2020, vol. 28), xvii–xviii.
Bone, Andrew G. "Introduction," in *The Collected Papers of Bertrand Russell*, ed. B. Russell (London – New York, Routledge, vol. 29, 2005), xlii–lii.
Boyer, Paul. *By the Bomb's Early Light: American Thought and Culture at the Dawn of the Atomic Age* (Chapel Hill, The University of North Carolina Press, 1994), 49–58.
Castelli, Alberto. *Una pace da costruire* (Milan, FrancoAngeli, 2002), 25–81.
Castelli, Alberto. *The Peace Discourse in Europe 1900–1945* (London – New York, Routledge, 2019), 21–37.
Holloway, David. *Stalin and the Bomb. The Soviet Union and Atomic Energy 1939–1956* (New Haven, Yale University Press, 1995), 306–308.
Ironside, Philippe. *The Social and Political Thought of Bertrand Russell* (Cambridge, Cambridge University Press, 1996), 85–88.
Killian, Report. https://history.state.gov/historicaldocuments/frus1955-57v19/d9.
Nehring, Holger. *Politics of Security British and West German Protest Movements and the Early Cold War, 1945–1970* (Oxford, Oxford University Press, 2013), 41–61.
Price, Matt. "Roots of Dissent. The Chicago Met Lab and the Origins of the Franck Report." *ISIS: The History of Science Society* 2 (June 1995), 224–244
Reiner, Braun, Robert Hinde, David Krieger, Harold Kroto, and Sally Milne. *Joseph Rotblat: Visionary for Peace* (Weinheim, Wiley VCH Verlag GmbH & Co., 2007).
Rempel, Richard A. "Introduction," in *The Collected Papers of Bertrand Russell*, eds. Bernd Frohmann, Mark Lippincott, Richard A. Rempel (London and New York, Routledge, 1988, vol. 13), xiv–xvi.
Roberts, Geoffrey. "Science, Peace and Internationalism: Frédéric Joliot-Curie, the World Federation of Scientific Workers and the Origins of the Pugwash Movement," in *Science, (Anti-)Communism and Diplomacy. The Pugwash Conferences on Science and World Affairs in the Early Cold War*, eds. Alison Kraft and Carola Sachse (Leiden – Boston, Brill, 2020), 43–79.
Rotblat, József. "Bertrand Russell and the Pugwash Movement. Personal Reminescences," http://digitalcommons.mcmaster.ca/viewcontent.cgi/context=russelljournal
Russell, Bertrand. "An Appeal to the Intellectuals of Europe (1915)," in *Justice in Wartime* (Chicago – London, The Open Court Publishing Co., 1916), 1–19.
Russell, Bertrand. "The Prospect of Permanent Peace," in *The Collected Papers of Bertrand Russell*, ed. Andrew G. Bone (1935, vol. 21), 79–87.
Russell, Bertrand. "How to Keep the Peace," in *The Collected Papers of Bertrand Russell*, ed. Andrew G. Bone (1935, vol. 21, 2008), 34–36.
Russell, Bertrand. *Which Way to Peace?* (London, Michael Joseph, 1936).

Russell, Bertrand. "Humanizing Warfare," in *The Collected Papers of Bertrand Russell*, ed. Andrew G. Bone (1937, vol. 21, 2008), 508–510.

Russell, Bertrand. "The Kind of Fear We Sorely Need," in *On Nationalism in The Collected Papers of Bertrand Russell*, ed. Andrew G. Bone (1950, vol. 26, 2020), 359–366.

Russell, Bertrand. *How Near Is War?* (London, Derrick Ridgway 1952), now in *The Collected Papers of Bertrand Russell*, vol. 26, 443–461.

Russell, Bertrand. "Western Values," in *The Collected Papers of Bertrand Russell*, ed. Andrew G. Bone (1952, vol. 26), 438–442.

Russell, Bertrand. "Man's Peril," in *The Collected Papers of Bertrand Russell*, ed. Andrew G. Bone (1954, vol. 28, 2003), 82–89.

Russell, Bertrand. "The Morality of 'Hydrogen' Politics," in *The Collected Papers of Bertrand Russell*, ed. Andrew G. Bone (1954, vol. 28, 2003), 49–53.

Russell, Bertrand, "The Hydrogen Bomb," in *The Collected Papers of Bertrand Russell*, ed. Andrew G. Bone (1954, vol. 28, 2003), 20–22.

Russell Bertrand. "The Road to Peace," in *The Bomb: Challenge and Answers*, ed. G. McAllister (London – Batsford, 1955), 47–68; now in *The Collected Papers of Bertrand Russell*, vol. 28, 352–355.

Russell, Bertrand. "Message to First Pugwash Conference," in *The Collected Papers of Bertrand Russell*, ed. Andrew G. Bone (1957, vol. 29, 2005), 341–345.

Russell, Bertrand. "Should H-bomb Tests Be Continued?" *The New Scientist*, March 28, 1957, now in *The Collected Papers of Bertrand Russell*, vol. 29, 310–315.

Russell, Bertrand. *Common Sense and nuclear Warfare* (London, Allen & Unwin, 1959, now also London – New York, Routledge, 2001), 31.

Russell Bertrand. *The Autobiography of Bertrand Russell* (London, Allen & Unwin, 1969, vol. III).

Russell Bertrand. *The Selected Letters of Bertrand Russell. The Public Years, 1914–1970*, ed. N. Griffin (London – New York, Routledge, 2001), 488–490.

Russell, Bertrand. "Britain Must Be Neutral (1935), Now Under the Title *How Not to Fight Fascism*," in *The Collected Papers of Bertrand Russell*, eds. E. Bone and M.D. Stevenson (London, Routledge, 2008, vol. 21), 37–39.

Russell, Bertrand. "Obstacles to World Government," in *The Collected Papers of Bertrand Russell*, ed. A.G. Bone (London – New York, Routledge, 2020, vol. 26), 46–61.

Ryan, Alan. *Bertrand Russell: A Political Life* (London, Penguin Books, 1988), 145–155.

Wang, Jessica. *American Science in An Age of Anxiety: Scientists, Anticommunism, and the Cold War* (Chapel Hill, University of North Carolina Press, 1999), 253–288.

Wittner, Lawrence S. *The Struggle against the Bomb* (Stanford, Stanford University Press, 1993), 269–293

Wittner, Lawrence S. *Confronting the Bomb: A Short History of the World Nuclear Disarmament Movement* (Stanford, Stanford University Press, 2009), 3–6.

2 Raymond Aron
International Relations in the Atomic Age

Francesco Raschi

Introduction

Raymond Aron, in his vast scholarly output, has among other things taken issue with the subject of war and international politics. In this regard, he was fond of repeating that war is "inseparable from politics."[1] His interest in these topics matured over the decades. Starting from decidedly pacifist positions in the late 1920s and early 1930s,[2] Aron then approached international issues since World War II.[3] As editor of *France Libre*, organ of the Gaullist resistance movement to the Vichy regime, he began to deal with both war in general and military strategy in particular (the topic that interests us here). After the war, he would continue analyzing international politics, along with international economics, as a columnist for the conservative-oriented French daily *Le Figaro*. Already at the end of the 1940s, he published a book that, in its first part, analyzed the planetary rivalry between the United States and the Soviet Union at first hand.[4] Three years later, in 1951, Aron offered a philosophical interpretation of the Thirty Years' War (1914–1945) and attempted to draw a picture of the Cold War and possible future scenarios, starting, on the one hand, from the downgrading of the (former) great European powers and, on the other, from the ideological and diplomatic rivalry between the United States and the USSR.[5] In the years that immediately followed, in addition to continuing to write weekly for *Le Figaro*, Aron also wrote a number of articles dealing with the theory of BRs and the method of investigating them.[6]

The great text, however, that gave Aron the status of a first-rate scholar of international relations, is the now classic *Peace and War*.[7] In this voluminous book, every issue concerning peace and war was treated in detail. The text was divided into four parts. In fact, Aron intended to investigate (war) conflict and international peacemaking by using a method that was at once philosophical, sociological, and historical. Only in this way, in his view, could a particular field of social action, such as international politics, be understood. In his view, there were four conceptual levels of understanding of that dimension. The theoretical one was necessary for the elaboration of the fundamental concepts and for a description of the typical situations of international politics (actors, international system, types of wars, and peace). The second level, i.e., the sociological,

DOI: 10.4324/9781003617174-3

investigated the circumstances that influenced the locations of conflicts between states (geography, material resources, population, etc.). The third level, the historical one, led Aron – and here, for our short contribution, the topic is most relevant – to investigate a precise diplomatic-strategic conjuncture, that of the Cold War. In this part of the book, Aron discussed the strategy of deterrence between the United States and the Soviet Union, by examining the relations between the two great powers and their allies within the blocs, as well as by analysing the role of the non-aligned countries in the context of the Cold War. He never concealed the implicit solidarity that bound the two "enemy brothers" in maintaining the *status quo*, including in the nuclear sphere, as we shall also see later.[8] Finally, the fourth level was the praxeological one, that is, the sphere in which the ethical-political dilemmas became pressing. This level was the one in which the antinomy between political action and moral action, which could never be resolved once and for all, confronted – both actors and commentators – the problem of problems in international politics, namely that of legitimate means on the one hand and universal peace on the other.[9]

Finally, still on the topics that interest us here, Aron would devote two volumes in the 1970s to analyzing the thought of the Prussian general von Clausewitz. In the first volume, he would have analyzed the general's thinking by treating *Of War* as a classic of political thought. Aron would come to regard von Clausewitz not so much as the militarist theorist of war but as the thinker who thought of war as the continuation of politics with the addition of other means. In the second volume, however, he would try to apply the categories developed by the Prussian general to the diplomatic-strategic conjuncture of the second half of the 20th century.[10]

At this point, however, it is necessary to go into the details of Aron's reflections on the change that the nuclear weapon had brought to international politics. For Aron, as will be seen, the atomic weapon assumed an indisputable centrality – it was the fundamental novelty of the new epoch! – in the international panorama after World War II. Having said this, the sociologist and philosopher believed, from the very first years after World War II, that not every political discourse, international or otherwise, could be traced back to the atomic question (and the mutual destruction that the use of such a weapon would cause). In fact, as we shall see more clearly in the remainder of this essay, Aron would continue to forcefully claim the autonomy of politics even in the nuclear age. And by autonomy of politics is to be understood precisely the possibility, on the one hand, of reaching compromises and appeasements and seeking a just political order, but also, on the other hand, the possibility of fighting conflicts that are not necessarily apocalyptic. In short, in this sense, even if the statement might appear provocative, the intellectual mark of Aron's contribution to relations in the shadow of the nuclear apocalypse seems to be that of wanting to save war. To be able to think of war as a limited and circumscribed event, which does not involve the inevitable climb to the extremes of Clausewitzian memory.

Impossible Peace and Improbable War

In 1948, in the text analyzing the great ideological schism between West and East, Aron formulated the diagnosis of the global rivalry between the two great powers that he would later use in many subsequent works – and which constitutes a slogan that is as effective journalistically as it is profound theoretically – which goes like this: "peace impossible, war improbable." Peace was impossible because, in a global diplomatic theater, the ideological conflict was now general: the contenders ultimately did not recognize each other, for doctrinal and other reasons. The Cold War, in fact, really questioned the limitation of what was at stake in the conflict between states. On the contrary, warlike peace was precisely characterized by the questioning of everything: from economic regimes to political systems, from spiritual and ideological beliefs to the survival of the ruling classes. At the same time, however, war was unlikely, because nuclear armaments seemed to steer global history towards a balance of terror, certainly precarious, but not necessarily short-lived.[11]

A few years later, in another text on the borderline between international relations theory and an essay on contemporary history and analysis of current affairs, Aron would analyze the diplomatic-strategic constellation of the Cold War, starting precisely from the analysis of the so-called Thirty Years' War (the two world wars).[12] In this text, with his usual analytical rigor, but with no less philosophical-historical depth, he reconstructed the Cold War in its ideological, economic, and military aspects.

One of the recurring questions about international politics that Aron tried to answer in those years was whether it was possible, i.e. still possible, to limit conflicts and wars in the atomic age. In this regard, Aron accepted neither the perspectives of the optimists who believed that the nuclear weapon, with its destructive effects, would necessarily lead to peace (war kills war), nor those of the pessimists who saw the nuclear apocalypse with the simultaneous end of civilization as inevitable. As a realist more by temperament than by conviction, he always believed that the atomic weapon had not reversed the course of international politics. War was – and let it be said in passing, had to be – even in the nuclear age, always possible, but it did not necessarily have to provoke escalation to extremes and mean communal suicide (the apocalypse).[13] Ultimately, for Aron, even conflicts in the nuclear age could, indeed had to, remain limited conflicts.

In the autumn of 1951, during the Korean War, Aron foresaw, after the intervention of the Chinese and also the stabilization of the front, that the conflict would end in a compromise of peace. Both powers(?) differently involved in the war, agreed on the final aim of the conflict, which was "negotiated peace":

> from the moment when, each side, for fear of an extension of the conflict, has given up hope of a decisive victory, the showdown must lead to a compromise ... The impossibility, recognised by both sides, of total victory logically leads to negotiated peace.[14]

In this sense, for Aron, already in the early 1950s, the Cold War was equivalent to a limited conflict in which each of the two camps used only a part of the means at its disposal.[15]

In another essay a few years later on the same subject, Aron questioned whether it was possible to limit war in the nuclear age. The problem was historically unprecedented since, in the atomic age, the threat of unlimited war depended not only on the fury of men, the nature of political regimes, and the size of the stakes, but also on the character of the weapons at hand. Well, for Aron, even in this era, it was not only possible but indeed necessary to limit warfare. For this to happen, it was necessary for the two major players to avoid the "scaling up to extremes" and not to set goals that they knew they could only achieve through the annihilation of the enemy. The states involved in the two world wars, for different reasons, as we know, were unable to limit the conflict. In World War I, the combination of the extension of the theater of operations and the logic of alliances coupled with industrialization – which provided a surplus of war material – and democratization – which provided an abundance of men through compulsory military service – led to the escalation to extremes. World War II, then, by introducing two further innovations, those of unconditional surrender and carpet-bombing, accelerated the debauchery of the conflict. The two nuclear weapons dropped on Japan, in this sense, were nothing but the almost inevitable development of the unconditional surrender of the enemy through the intensive bombing of cities.[16]

From 1946 to 1949, the Americans used their nuclear monopoly in an exclusively defensive manner: they did not, according to Aron, force the Soviets' hand in any way (e.g., they did not force the USSR to lift the Berlin blockade). On the contrary, in some ways, they took advantage of this situation to reduce the conventional military budget and create a kind of balance: their ability to devastate Soviet cities was offset by the Soviet ability to militarily occupy Western Europe with conventional weapons.

Korea witnessed the first successful attempt to limit the conflict, both geographically and in terms of warfare. On the first front, the Americans resigned themselves, according to Aron, to limiting their operations to Korean territory only. On the second, however, they excluded the use of nuclear weapons. In addition to these two limitations, there is another, that of target limitation:

> General MacArthur's formula – there is no alternative to victory – is succeeded by the not openly formulated, but implicitly admitted conception: only negotiated peace on the basis of a balance of forces makes it possible to avoid the extension of hostilities.[17]

In short, it was only possible to limit the war, even in the nuclear age, if one clearly renounced both unconditional surrender and capitulation by the enemy. Renouncing total victory and capitulation did not, however, exclude the possibility of battlefield successes that forced the enemy to negotiate.

So, was the use of nuclear weapons to be ruled out? Aron is never so categorical on any subject, let alone this one. However, he considered a nuclear conflict unlikely. Even in the early 1950s, he did not think that the use of atomic weapons would necessarily lead to a general and total war. In his view, at least on a hypothetical level, a military aggression by a state against its neighbors could involve the use of atomic warheads, even if only to disperse invading troops or destroy their bases. Thus, even the use of tactical nuclear weapons, in Aron's opinion, would not necessarily have triggered a worldwide conflict, and especially not necessarily have led to the mutual annihilation of the two contenders.[18] Even the eventual atomic war was – and should, if possible, remain – an instrument in the hands of politics. An eventual conflict in Europe for Aron would most likely have been atomic, but as already mentioned, it would not necessarily have led to the mutual self-destruction of the contenders. The abundance of nuclear armaments, as well as the means of transporting them, made decisive success unlikely even for whoever took the initiative first. Thus, atomic equality would, for the umpteenth time in history, make "enemy armies the number one target."[19]

What, then, was to be done to prevent the destruction of cities with atomic weapons? Westerners, for Aron, had at least two means to avoid the destruction of their (and others') cities. The first means was to solemnly proclaim that they would not take the atomic initiative, but that at the same time they would respond harshly – "ruthlessly" – to the enemy's initiatives. The second instrument, on the other hand, stipulated that even in the event of a general war they should refrain from pursuing objectives that were incompatible with the survival and dignity of the enemy states: "it would be wise not to aim for unconditional surrender, even in the case of a general war."[20] In short, Aron already criticized in the early 1950s the postulate that any crossing of the nuclear threshold would necessarily lead to the extreme of total nuclear war.[21]

Solidarity between the United States and the Soviet Union in the Age of Deterrence

Aron, as mentioned, participated as one of the leading French experts in the debates on strategy in the nuclear age. As is well known, the US atomic monopoly ended in 1949, while US superiority in intercontinental missiles ended in 1957, when the Soviets also acquired the ability to strike US cities with intercontinental missiles.

Aron, on more than one occasion, would rightly note how, even in the era of American nuclear superpower, the Soviets were in fact never paralyzed or terrified by the undoubted capacity of the United States to strike and devastate Russian cities:

In spite of everything, at no moment between 1945 and 1957 did the Soviet Union seem paralysed or terrified by the capacity which American strategic aircraft certainly possessed to devastate its cities. The course of the Chinese civil war was not affected by it. The atomic threat neither halted North Korean aggression nor prevented the Chinese intervention nor hastened the conclusion of an armistice. The changes in style of Soviet diplomatic strategy after 1953 are manifestly attributable to Stalin's death, to the disputes and personalities of his successors, not to modifications in the relations of atomic or thermonuclear forces.[22]

This is essential to understand the Aronian approach to strategy in the atomic age. His approach was remarkably bereft of categorical judgments and, most crucially, it could not be ascribed to one single cause, even if it was the power of the means of destruction. In short, for Aron, even during the Cold War era (or belligerent peace, as he liked to repeat) many factors other than the nuclear weapon came into play; factors that were both economic and ideological, and finally psychological. Ultimately, he believed that politics continued to have its own margin of autonomy and over-ordination even in these stormy times.

The mechanism of deterrence, Aron often repeated, did not originate with the atomic age. Such a mechanism had always existed and depended mainly on the material means available to one state to stop another, as well as the resolve to use those means if necessary. In the atomic age, the only novelty of deterrence depended, if anything, on the material consequences of carrying out the threat. The strategy of deterrence, wrote the French sociologist and philosopher, was "essentially a test a will power, an exchange of alternate threats and messages, or rather of threats bearing messages and messages pregnant with threats."[23] If the threat of war, even of the thermonuclear war, he continued, were to be part of current international relations, it could hardly be ignored, as it would appear to the main actors as contrary to all rationality. This is why, Aron wrote, the threat of thermonuclear war brought with it

> the connivence of enemies as well as for the suspicions among allies; for the impossibility of reconciling the enemies and of continued trust among allies; for disarmament talks as well as the arms race; and for the obsessional fear of war coupled with both sides are arming will never have to be fought.[24]

Thus, a first effect caused by nuclear weapons seemed to be the interest of the two big ("duopolists") in maintaining the status quo and thus, in a sense, solidarizing with each other.[25] This solidarity consisted in the fact that the "duopolists" would continue to have an interest in not destroying each other (favoring the victory of third states) and also in preventing the spread – this argument would have served Aron well to frame historically and thus to understand the hostile attitude of the United States towards the French *force de frappe* – of armaments to other states.

In *Peace and War among Nations,* Aron would analyze what he saw as the different models of nuclear conflict (and deterrence) in the early 1960s. On a theoretical level, if only two powers had held the bomb, two different scenarios could have been envisaged. The first was the so-called "impunity of the crime" (or also called the "two gangsters" situation); the second, vice versa, defined from the equality of crime and punishment.[26]

Let us start with the first. If one of the two big guys, by firing the first shot, had been able to benefit from impunity by virtue of the destructive capacity of the reprisal forces, then we would have been faced with the situation of the confrontation between the two gangsters: whoever fired first, hitting the target, would have won the confrontation once and for all.

The second scenario was equally simple and straightforward: if those who had suffered the first blow were able to retaliate, we would be faced with a kind of equality – perhaps paralyzing equality – between crime and punishment.

In Aron's view, therefore, the whole issue depended on whether the thermonuclear apparatus was vulnerable. While the vulnerability of the apparatus created an inherently unstable situation that could result in a generalized and perhaps even total war because leaving the first blow to the adversary could have had irreparable effects, the full invulnerability of the apparatus, on the other hand, would inevitably lead to a situation of equilibrium (i.e., equivalence of damage) that seemed, ipso facto, to eliminate the very idea of victory. Precisely in this regard, Aron asked himself, rhetorically but also provocatively, whether it made sense to fight a war that could in no way be won.

The models outlined by Aron, however, were abstract ones, in some ways ideal types in the proper Weberian sense of the term, i.e., boundary concepts that served to measure reality. In his view, they had precisely little chance of being realized. In history, in fact, the most likely model could have been the intermediate one, namely what Aron christened "the inequality of the crime and the punishment." In the event of a war between the United States and the USSR, it would have been unlikely that the first strike would have eliminated the other's means of retaliation: "But it is also improbable that the superpower on whose territory several dozen thermonuclear bombs of five or six megatons have fallen would have been in a position to inflict a punishment in proportion to the crime [...] Let us suppose, in effect, that each of the duopolists knows that if he strikes first, he will endure three times less damage than if he had left the initiative to his adversary. In other words, relative victory would go to the power dealing the first blow, and neither of the two powers is unaware that this is the case. Of course, even the victor would be severely damaged and would prefer non-war to relative victory, if he had the choice. But he would also prefer relative victory to relative defeat [...] In other words, in abstract terms, every situation of inequality of the crime and the punishment, particularly if this inequality is imputable to the vulnerability of the thermonuclear system, creates the risk of what the American authors call the "pre-emptive strike, the blow dealt in anticipation not the blow one side suspects the other to be on the point of dealing."[27] Here, Aron drew a clear distinction between "Pre-emptive War" and "Preventive War." The "Preventive War" was a war waged in cold blood, at a time that was deemed favorable by a state. The "Pre-emptive War," on the other hand, was unleashed in times of crisis, not because victory was preferred to peace, but because one expected to be attacked at any moment.[28]

As said, the scenario of destroying all the enemy's means of retaliation was quite improbable. According to Aron, in fact, neither Major would ever have had the capacity to destroy all the other's means of retaliation in one fell swoop.[29] Hence, the further one moved away from the "two gangsters" situation, the more the temptation to take the initiative and strike first diminished. In such a scenario, the greatest danger was the *escalation* of a secondary conflict, which gradually led to some form of *escalation*. In such a scenario, in fact, the fact that the nuclear threshold had been raised – only for the vital purposes of the two superpowers – risked the loss of the weapon's deterrent role for posts that did not jeopardize the survival of the United States and USSR.

In any case, the vulnerability of the thermonuclear apparatus was precisely the factor that increased, or could increase, the risk of *escalation*. If, on the contrary, neither of the duopolists had the means to destroy or strike the enemy's thermonuclear apparatus to death, then the obsession with the other taking over would slowly fade away. According to Aron, therefore, the climb to the extremes – the nightmare of mankind in those more or less dark times – became more and more improbable:

> In other words, the stability towards which relations between the duopolists tend, as the punishment approaches equality with the crime, excludes the extension of the secondary conflicts and simultaneously renders the thermonuclear apocalypse more unlikely and limited wars more likely. Monstrous war and the effectiveness of the threat of such a war against any provocation occurring at the same time is inconceivable.[30]

Although it seemed legitimate to affirm that, in abstract terms, the relative invulnerability of thermonuclear apparatuses combined with the approximate equality of "crime and punishment" increased the probability of limited conflicts and not of total war, Aron did not (not even in this case) abandon his well-known predisposition to systematic doubt. In fact, he readily admitted that the nuclear deterrence mechanism itself had an aleatory and uncertain character, given that, in the matter of nuclear weapons, there were no historical examples other than dated ones (the detonation of the two devices on Hiroshima and Nagasaki was, in some ways, part of another historical epoch, completely different from the present), but only calculations and laboratory exercises. A second reason was that the political decision-makers described in this modeling were abstract, whereas the political leaders who had their "finger on the button" were instead flesh-and-blood beings, who had their own perception of reality, a certain ideology, certain material interests, and particular criteria of judgment that did not make them easily adhere to a rigid formal rationality.[31]

Massive Reprisals and Flexible Response

As it has already been said, Aron had participated in Western debates on nuclear strategy since the early 1950s, with a leading role. As is well known, the Eisenhower administration, which came into office during the last phase of the Korean War, undertook a review of American nuclear strategy. In 1954, Secretary of State Foster Dulles presented the so-called "*massive re*taliation" doctrine. If, up to that time, the US preferred to source the enemy's various aggressions with direct and local resistance, with the new doctrine they reserved the right to intervene with powerful instruments of instant retaliation – not excluding the atomic weapon – in the direction of targets decided by them at their discretion. And this retaliation, of course, also applied to aggression outside American territory. This doctrine was obviously much appreciated by the Europeans because it constituted a concrete sign – one might almost say a solemn commitment – on the American side to defend its allies by threatening the Soviets with the use of the ultimate weapon.

Aron pointed out the problematic nature of this strategy early on in his articles on international politics in *Le Figaro*. Aron said he understood the reasons for Eisenhower's administration – reasons that were mostly dictated by budgetary needs (concentrating all development on weapons of mass destruction, reducing conventional armaments) and internal consensus (reducing, in perspective, the number of American soldiers deployed in Europe and Asia). That said, Aron considered this strategy to be problematic because, in his view, the atomic weapon could not be wielded on a continuous basis. In other words, the threat of massive retaliation in the face of every local aggression could in no way be deemed credible by the enemy. In some ways, indeed, the subordination of all military means to the atomic weapon risked not being very effective and was certainly less productive than American politicians claimed. Such a strategy, which Aron defined as "all or nothing," originally aimed at increasing the margins of American strategic freedom, on the contrary risked restricting it considerably, putting American leaders faced with the not-so-rosy prospect, in the face of Soviet aggression towards their allies, between "capitulation and general war."[32]

In 1955, while in the United States as a correspondent for *Le Figaro*, Aron read an article by a US military man, a certain Richard Leghorn, who frontally criticized the US administration's nuclear strategy.[33] Aron was persuaded by this essay. In fact, in his opinion, Leghorn successfully tried to show that war, even atomic war, was not necessarily going to assume the apocalyptic character so terribly imagined by the pessimists. Reading this article, which, in addition to criticizing the doctrine of massive replication also called for the use of tactical nuclear weapons with weaker explosive charges, definitively persuaded Aron: even an open conflict between the two great powers would not necessarily result in nuclear apocalypse. In short, politics, reasonable politics in particular, could have continued to be architectural, i.e. it could have continued to fulfill its purpose, which was to moderate or limit the conflict.

Much more congenial, from this point of view, would have appeared to Aron the doctrine developed by the Kennedy administration in the early 1960s (and already anticipated by Leghorn). As is well known, the strategic parity and then the *missile gap* – more imagined than real, in hindsight – led the Kennedy administration to radically change strategy. The *massive retaliation* – appropriate at a time when there was a clear disparity between the capabilities of the two powers – was succeeded by the *flexible response*, which envisaged a response commensurate with the degree of the offense: in short, the strategy of responding with atomic weapons was abandoned – the all-or-nothing strategy, opposed by Aron because it forced a choice between going to extremes (use of the thermonuclear weapon) and passivity – in favor of a response characterized by different stages.

The "flexible response" could not, therefore, fail to please Aron. The reason can easily be understood: this strategy went precisely in the direction our author had been hoping for at least a decade, that of conflict limitation. Limitation that consisted in minimizing the risk of escalation to extremes. In what way? By always deferring the use of the atomic weapon as much as possible in the same strategic documents. Thus, war, even in the atomic age, continued to be an instrument of politics, perhaps one of the main instruments, if not the main instrument, of international politics.

In the early 1960s, Aron played a leading role in both the scientific and political spheres. In the latter sphere, in fact, he played a decisive role in socializing both public opinion and the French and European ruling classes with regard to the redefinition of American strategic coordinates and also, above all, in preventing misunderstandings from arising within the Atlantic alliance on the American desire to protect the Old Continent.[34] It is well known, in fact, how originally the "flexible response" was interpreted in Europe as a warning of American disengagement from the European continent, in favor of safeguarding US territory ("sanctuary"). In short, there was a widespread fear that the Old Continent, in the American view, was becoming a potential battleground for new wars. Aron thought otherwise. European anxieties, in his view, depended on a misunderstanding of the American strategic debate. The McNamara doctrine had, in fact, the primary purpose of avoiding the alternative, which Aron calls "oversimplified and lethal,"[35] between all-or-nothing (that is, between scaling to extremes and total war, or between passivity in the face of enemy threats or aggression). Graduated retaliation, in fact, envisaged that in the event of provocations one could respond first with conventional weapons[36] and, in the case of the use of nuclear weapons, before counterforce (that is, of the indiscriminate attack against the population living in the cities, and which could have been a true foretaste of the nuclear apocalypse). Although, in this sphere, that of crossing the nuclear threshold, anything was possible because one entered, whether one wanted to or not, into a "beyond which we enter a vast unknown."[37]

Instead of distrusting America and the new American strategic doctrine, the European allies should have convinced themselves that

> A nation that has at its disposal a wide range of possible responses, rather than a narrow choice between thermonuclear apocalypse and capitulation, is far more likely to inspire caution and respect in potential aggressors, including one contemplating more local operations. In other words, an increase in conventional arms and the ability to conduct combat operations without having to resort to nuclear weapons should also be regarded as deterrents, indeed as the means of deterring local aggression against which the threat of massive retaliation might prove futile.[38]

The American position, as mentioned, convinced Aron because it was a "reasonable response," a sort of compromise that, like all compromises, never fully satisfied everyone but constituted a sort of lesser evil capable of postponing "the most monstrous of dangers" as much as possible. The flexible response, after all, for Aron expressed precisely

> the contradiction contained in the concept of escalation, a process simultaneously feared and exploited as a threat. It is this very contradiction that defines the essential nature of the thermonuclear duel: the opponents strive to obtain certain results by making threats they do not intend to carry out and, unable to agree to drop their threats, try to go on playing the game.[39]

The strategy of graduated response, as mentioned above, seemed reasonable to Aron because it distinguished between the use of various types of weapons; because in choosing between various dangers it distinguished between initial confrontations and extreme weapons; and because it was finally a strategy that avoided escalation to extremes by accident, without, however, completely excluding this scenario, which was still possible:

> If no distinctions are made between initial operations and ultimate weapons, there is a risk of ultimate escalation by accident or misunderstanding. In short, escalation is at once a danger that needs to be met and a threat that could not and should not be surrendered.[40]

Thus, if the graduated retaliation represented an acceptable compromise between different demands, it is equally true that it was not a model that by virtue of its stringent logic induced unconditional adherence.[41]

With his usual lucidity, Aron did not hide the reasons for a certain European anxiety – in part, therefore, justified – towards this new strategy. The Europeans, in fact, were faced with two indisputable facts: (i) their security was no longer based, as it had been in the days of the Eisenhower administration, on the certainty of the United States' atomic replication; (ii) the disparity of dangers to which European and American territory was exposed:

> But it took the McNamara doctrine to force Europeans into realising that their situation was ultimately not identical with that of the United States and that there was less truth than wishful thinking in the old bromide about all of us being in the same boat.[42]

But this difference in condition, due to the fact that European cities were within range of Soviet medium-range missiles, did not, according to Aron, depend so much on the MacNamara doctrine or on the ill-will of the Americans, as on "geography":

> Actually this state of affairs should be blamed on geography rather than on the President of the United States; for even in the thermonuclear age distance has not been entirely eliminated as a factor, except in the imagination of so-called experts inclined to confuse strategy with science fiction. Europe's proximity to the presumptive enemy entails increased vulnerability to bombers and missiles. Continental Europe and the Soviet empire share the same land mass, and the common frontier implies at least a theoretical possibility of local aggression with conventional weapons below the atomic threshold.[43]

Geography, in short, could actually make the "flexible response" more favorable to US interests and less to European ones. This could legitimately give rise to a climate of mutual distrust between the Europeans and the Americans – distrust that was obviously then accentuated by the two corollaries of the graduated response: the American decision to maintain a monopoly on the one hand and the centralization of nuclear arms control on the other.

This US decision, as is well known, was strongly opposed by the French ally. In particular, the president of the Republic, de Gaulle, is said to have contested the Kennedy administration's new doctrine and, in opposition to it, decided to carry out, independently of any coordination with the allies, a nuclear *force de frappe* as an indispensable element of national defense. Aron, who was not entirely insensitive to Gaullist arguments[44] and especially blamed certain American rigidities for France's angry reaction,[45] was at the same time critical of the Gaullist position.[46] This is not the place to go into these aspects in depth, but at least a mention of Aron's criticism of de Gaulle's foreign policy is necessary. The Atlantic alliance, in Aron's view, was essential, and so was coordination within it. The *force de frappe*, in his view, would never have been able to guarantee an autonomous deterrence capability and could at best have been a complementary force to the American one.[47] What Aron mainly reproached de Gaulle for, however, was not so much the atomic autonomy as the "diplomacy of the fist on the table, of the scandal."[48] In his opinion, in fact, the General's press conferences – the one in which he vetoed British entry into the EEC or in which he disdainfully rejected the proposal to equip French submarines with Polaris missiles, provided they were integrated into NATO – although they were "works of art," were at the same time true "historical-political acrobatics" which, if emulated, could have caused resounding damage to the western camp. In this regard, he writes this: "I think the Atlantic alliance is solid enough to afford the luxury of a great man, careful of his own stature and secrecy. But if unfortunately, the great man found imitators, the alliance would not endure."[49]

For Aron, de Gaulle was not naive enough to really believe that the French nuclear force could be an effective weapon of deterrence towards the Soviet Union (the presence of American soldiers in Germany was much more so!). But the General's work was open to criticism because it nevertheless constituted "bad pedagogy" which, taking advantage of a latent anti-Americanism present in French public opinion, used dangerous rhetoric that, by pitting "Europeans" against "Atlanticists," ultimately played into the hands of the Soviets.[50]

Conclusions

It is, however, time to end our discourse. As we have seen, Raymond Aron was convinced that first the nuclear weapon (and then the thermonuclear weapon) was a major innovation in international politics. But at the same time, he was equally convinced that the terribleness of a nuclear war was, at least rationally, to be ruled out. Thermonuclear war, in short, would have been sheer madness. In a way, he was convinced that from the great evil of the extreme weapon (a total war) a lesser evil (a limited war) could arise. If it is true, as Aron writes in the chapter on war in *Espoir et peur du siècle* that what "we must avoid is the generalisation of a conflict," then the nuclear weapon could indeed help to avoid such generalisation. At the same time, however, the risk of madness "increases if we prepare for only one war, the one we do not want to fight, if we close ourselves off, through a senseless military policy, in the all or nothing alternative of apocalypse or capitulation."[51] What was therefore also needed in the nuclear age, in Aron's opinion, was for politics, understood in a Clausewitzian way, as the intelligence of the state, not to fail even at the outbreak of hostilities:

The spirit of reasoned intent that informs policy must not be allowed to evaporate the moment the first bombs start exploding; intelligent national policy must to the very end make a determined effort simultaneously to safeguard the national interest and to prevent escalation to the extremes of violence.[52]

In conclusion, it was evident that, in the Aronian perspective, war could most probably not be expunged from history once and for all; one could – one should! – however, try to limit the volume of conflicts. In such a context, it was therefore desirable for states to be guided by prudence in their dealings.

Even in the nuclear age, therefore, international relations remained social relations that sometimes required recourse, therefore always possible and even legitimate, to force. Precisely for this reason, in Aron's view, the morality of wisdom, as it has been defined by some interpreters,[53] was an attempt to safeguard politics and its autonomy from the attempts to override it inherent in linear and mono-causal approaches. In this regard, Aron criticized the so-called idealist illusion that too often risked slipping into the side of fanaticism. Fanaticism precisely in the sense of dividing states into good and bad and wanting at all costs to punish the latter not only militarily but also morally and legally. As he wrote, the idealists very often believed they were breaking with power politics by exaggerating the crimes, because in their attempt to replace politics (and thus also the eventuality of war) with law (and the morality of law), they overlooked the fact that there were dissatisfied and revolutionary states that had to be contained through balance and even power politics, and not through moral condemnation and political annihilation. At the same time, however, he also reproached the so-called realists who thought of international relations only in terms of *Machtpolitik*: they, too, had to realize that the realm of politics involves conflict, but not only conflict, and that political ideologies, the pursuit of a fair political order, have, in turn, their own importance in diplomatic-strategic conduct. Hence, in his view, even in the shadow of the nuclear apocalypse, the traditional, antinomian morality of diplomatic-strategic action had not failed. The best course of action, in short, was – continued to be – the prudential one. In this respect, Aron defines prudence in international politics as follows: "To be prudent is to act in accordance with the particular situation and the concrete data," and not in accordance with some system or out of passive obedience to a norm or pseudo-norm; it is to prefer the limitation of violence to the punishment of the presumably guilty party or to a so-called absolute justice; it is to establish concrete accessible objectives conforming to the secular law of international relations and not to limitless and perhaps meaningless objectives, such as "a world safe for democracy" or "a world from which power politics will have disappeared."[54] One could not have escaped from "warlike history," but at the same time one should not have "betrayed ideals" either.

Aron concluded his text on *Peace and War* – the title page of which featured as an exergue a quotation, which was nothing but praise for moderation and prudence, from his beloved Montesquieu[55] – in this way:

> Let us leave to others with more talent for illusions the privilege of speculating on the conclusion of the adventure, and let us try not to fail either of the obligations ordained for each of us: not to run away from a belligerent history, not to betray the ideal; to think and to act with the firm intention that the absence of war will be prolonged until the day when peace has become possible – supposing it ever will.[56]

Notes

1. Raymond Aron, *Introduction à, Max Weber, Le Savant et la politique* (Plon, 1992), 43.
2. See Francesco Raschi, "L'addio di Raymond Aron al pacifismo," *Politics* 13, no. 1 (2020), 27–44.
3. See Aron's indispensable intellectual biography for a general overview of the author and his thought, Nicolas Bavarez, *Raymond Aron. Un moraliste au temps des idéologies* (Flammarion, 1993). Another comprehensive work in which a number of authors have analyzed the multifaceted work of the French philosopher and sociologist is the following: José Colen and Elisabeth Dutartre, *The Companion to Raymond Aron* (Palgrave Macmillan, 2015). For another introduction to Raymond Aron's political thought, including international thought, see Gwendal Châton, *Introduction à Raymond Aron* (La Découverte). Also, have a look at the must-haves: Pierre Hassner, "Raymond Aron et la philosophie des relations internationales," *Commentaire* 122 (Été 2008), 638–642; Stanley Hoffman, "Raymond Aron and the Theory of International Relations," *International Studies Quarterly* 29, no. 1 (1985), 13–27.
4. Raymond Aron, *Le grand schisme* (Gallimard, 1948).
5. Raymond Aron, *The Century of Total War* (Beacon Press, 1966). This text is a translation of Raymond Aron, *Les guerres en chaîne* (Gallimard, 1951).
6. All these writings are now included in Raymond Aron, *Les sociétés modernes* (Presses Universitaires de France, 2005). In English, for some of the most important writings on international relations, see also Raymond Aron, *Politics and History* (Transaction Books, 1984), 102–121, 166–185, 186–211, 212–236.
7. Raymond Aron, *Peace and War. A Theory of International Relations* (Routledge, 2017).
8. Aron, *Peace and War*, 615–652.
9. On these topics, the publication in Italian – a world premiere – of a course by Aron, given in 1973, on political action and its antinomies is of fresh interest. See Raymond Aron, *Teoria dell'azione politica* (Marsilio, 2023).
10. Raymond Aron, *Penser la guerre, Clausewitz. L'Âge éuropèenne* (Gallimard, 1976); Raymond Aron, *Penser la guerre, Clausewitz. II. L'Âge planétarie* (Gallimard, 1976). An introduction to Aron's interpretation of Clausewitz see Robert Colquhoun, *Raymond Aron. The Sociologist in Society 1955–1983*, vol. II (SAGE Publications, 1986), 447–468.
11. Aron, *Le grande schisme*, 26–31.
12. See Aron, *The Century of Total War*, 74–91 and 169–180.
13. Raymond Aron, *Espoir et peur du siècle. Essais non partisans* (Calmann-Lévy, 1957).

14 Raymond Aron, *De la paix sans victoire. Note sur les relations de la stratégie et de la politique* (1951), in Aron. *Les sociétés moderes*, 949.
15 Aron, *De la paix sans victoire*, 949–950.
16 Raymond Aron, *À l'âge atomique peut-on limiter la guerre?* (1955); Aron, *Les societés modernes*, 971–985.
17 Aron, *À l'âge atomique peut-on limiter la guerre ?*, 975.
18 Aron, *À l'âge atomique peut-on limiter la guerre?*, 979.
19 Aron, *À l'âge atomique peut-on limiter la guerre?*, 981.
20 Aron, *À l'âge atomique peut-on limiter la guerre?*, 981.
21 Cf. Christian Malis, *Raymond Aron et le débat strategique français 1930–1966* (Economica, 2005); and also Raymond. Aron, *Memoirs. Fifty Years of Political Reflection* (Holmes & Meier, 1990). In particular, see the 11th chapter entitled: *The Wars of the Twentieth Century*.
22 Aron, *Peace and War*, 421–427.
23 Raymond Aron, *The Great Debate. Theories of Nuclear Strategy* (Doubleday & Company, 1965), 223.
24 Aron, *The Great Debate*, 226–227.
25 "In a thermonuclear duopoly the super powers have a double interest in common: not to destroy each other (and thereby assure the victory of third powers), not to favour and, if possible, to prevent the dissemination of the decisive and terrifying weapons. For ten years it has seemed as if the two super powers (especially the United States) were at every moment conscious that their common interest in avoiding war prevailed over their opposed interests, however important; as if they were equally concerned to delay the moment when the accession of France and China to the thermonuclear club would put an end to the duopoly. The Soviet Union, despite socialist solidarity, has not helped China any more than the United States, despite the Atlantic Pact, has helped France. Neither alliances nor hostilities have ever been total, down through history. Solidarities among enemies, oppositions among allies assume an original form in the thermonuclear age" (Aron, *Peace and War*, 428).
26 Aron, *Peace and War*, 404–440.
27 Aron, *Peace and War*, 410–411.
28 Aron, *Peace and War*, 411.
29 Aron, *Peace and War*, 412.
30 Aron, *Peace and War*, 413.
31 Aron, *Peace and War*, 434–440.
32 Raymond Aron, *Les articles du Figaro. Tome I. La Guerre Froide 1947–1955* (Éditions de Fallois, 1990), 1186–1191.
33 Richard Leghorn, "No Need to Bomb the Cities to Win Wars: A New Counter-Strategy for Air Warfare," *US News & World Report*, January 28, 1955, 78–83.
34 On these aspects, see Malis, *Raymond Aron et le débat stratégique françai*, cit., 535–699; but also Massimo Guareschi, *I volti di Marte. Raymond Aron Sociologist and Theorist of War* (OmbreCorte, 2011), 63–103.
35 Aron, *The Great Debate*, 67.
36 From this point of view, it is, for Aron, more than understandable that the Americans call on the Europeans to carry out a major conventional rearmament.
37 Aron, *The Great Debate*, 74.
38 Aron, *The Great Debate*, 68.
39 Aron, *The Great Debate*, 217.
40 Aron, *The Great Debate*, 216.
41 Guareschi, *L'ombra di Marte*, 81.
42 Aron, *The Great Debate*, 88.
43 Aron, *The Great Debate*, 89.

44 Over the years, he has never categorically ruled out the possibility that France should acquire atomic armaments.
45 See Malis, *Raymond Aron et le débat stratégique français*.
46 On these topics see Lucia Bonfreschi, *Raymond Aron e il gollismo 1940–1969* (Rubbettino, 2014), 396–460; and may I also refer to Francesco Raschi, "Prospettive euroscettiche: critiche all'Europa nella storia dell'integrazione," in *Prospettive euroscettiche: critiche all'Europa nella storia dell'integrazione* (Editoriale Scientifica, 2020), 163–181. See also Elisa Piras and Francesco Raschi, "A Realistic Scepticism: Raymond Aron's Perspective on the European Construction," *Journal of European Integration History* 26 no. 2 (2020), 267–284.
47 Aron, *The Great Debate*, 105.
48 Aron, *Memoirs*, 434.
49 Raymond Aron, *Les articles du Figaro. Tome II. La Coexistence 1955–1965* (Éditions de Fallois, 1990), 1138–1141.
50 Aron, *Memoirs*, 446.
51 Aron, *Espoir et peur du siècle. Essais non partisans*.
52 Aron, *The Great Debate*, 80.
53 See D. J. Mahoney, *The Liberal Political Science of Raymond Aron* (Rowman & Littlefield Publishers, 1992). On Aron's political thought see Francesco Raschi, *Raymond Aron e le ideologie del Ventesimo secolo. Un liberale tra destra e sinistra* (MUP, 2005); Brian C. Anderson, *Raymond Aron: The Recovery of Political* (Rowman & Littlefield, 1997). See also to work critically on Raymond Aron's international thought Reed M. Davis, *A Politics of Understanding. The International Thought of Raymond Aron* (Louisiana University Press, 2009), 85–133 and 165–180 and on moderate political thought the recent Joshua L. Cherniss, *Liberal in Dark Times. The Liberal Ethos in the Twentieth Century* (Princeton University Press, 2021), 102–136.
54 Aron, *Peace and War*, 585.
55 The law of nations is naturally founded on this principle, that different nations ought in time of peace to do one another all the good they can, and in time of war as little injury as possible, without prejudicing their real interests (Charles de Secondat Montesquieu, *The Spirit of the Law*, I.3)
56 Aron, *Peace and War*, 787.

Bibliography

Anderson, Brian C. *Raymond Aron: The Recovery of Political* (Rowman & Littlefield, 1997).
Aron, Raymond. *Le grand schisme* (Gallimard, 1948).
Aron, Raymond. "De la paix sans victoire. Note sur les relations de la stratégie et de la politique," in *Les societés modernes* (1951), 939–952.
Aron, Raymond. *Les guerres en chaîne* (Gallimard, 1951).
Aron, Raymond. "À l'âge atomique peut-on limiter la guerre?" in *Les societés modernes*, 971–985.
Aron, Raymond. *Espoir et peur du siècle. Essais non partisans* (Calmann-Lévy, 1957).
Aron, Raymond. *The Great Debate. Theories of Nuclear Strategy* (Doubleday & Company, 1965), *The Century of Total War* (Beacon Press, 1966).
Aron, Raymond. *Penser la guerre, Clausewitz. L'Âge éuropèenne* (Gallimard, 1976).
Aron, Raymond. *Penser la guerre, Clausewitz. II. L'Âge planétarie* (Gallimard, 1976).
Aron, Raymond. *Politics and History* (Transaction Books, 1984).
Aron, Raymond. *Les articles du Figaro. Tome I. La Guerre Froide 1947–1955* (Éditions de Fallois, 1990).

Aron, Raymond. *Les articles du Figaro. Tome II. La Coexistence 1955–1965* (Éditions de Fallois, 1990).
Aron, Raymond. *Memoirs. Fifty Years of Political Reflection* (Holmes & Meier, 1990).
Aron, Raymond. *Introduction à, Weber, Max. Le Savant et la politique* (Plon, 1992).
Aron, Raymond. *Peace and War. A Theory of International Relations* (Routledge, 2017).
Aron, Raymond. *Les societés modernes* (Presses Univeristaires de France, 2025).
Bavarez, Nicolas. *Raymond Aron. Un moraliste au temps des ideologies* (Flammarion, 1993).
Bonfreschi, Lucia. *Raymond Aron e il gollismo 1940–1969* (Rubbettino, 2014).
Châton, Gwendal. *Introduction à Raymond Aron* (La Découverte, 2017).
Cherniss, Joshua L. *Liberal in Dark Times. The Liberal Ethos in the Twentieth Century* (Princeton University Press, 2021), 102–136.
Colen, José and Dutartre, Elisabeth. *The Companion to Raymond Aron* (Palgrave Macmillan, 2015).
Colquhoun, Robert. *Raymond Aron: The Sociologist in Society 1955–1983* (SAGE Publications, 1986).
Davis, Reed M. *A Politics of Understanding: The International Thought of Raymond Aron* (Louisiana University Press, 2009).
Guareschi, Massimiliano. *I volti di Marte. Raymond Aron Sociologist and Theorist of War* (OmbreCorte, 2011).
Hassner, Pierre. "Raymond Aron et la philosophie des relations internationales," *Commentaire* 122 (Été 2008), 638–642.
Hoffman, Stanley. "Raymond Aron and the Theory of International Relations," *International Studies Quarterly* 29, no. 1 (1985), 13–27.
Leghorn, Richard. "No Need to Bomb the Cities to Win Wars: A New Counter-Strategy for Air Warfare," *US News & World Report*, January 28, 1955, 78–83.
Mahoney, Daniel J. *The Liberal Political Science of Raymond Aron* (Rowman & Littlefield Publishers, 1992).
Malis, Christian. *Raymond Aron et le débat stratégique français 1930–1966* (Economica, 2005).
Meszaros, Thomas, and Antony Dabila. "Raymond Aron's Heritage for the International Relations Discipline: The French School of Sociological Liberalism," in *Raymond Aron and International Relations*, ed. Olivier Schmitt (Routledge, 2019), 142–162.
Raschi, Francesco. *Raymond Aron e le ideologie del Ventesimo secolo. Un liberale tra destra e sinistra* (MUP, 2005).
Raschi, Francesco. "L'addio di Raymond Aron al pacifismo," *Politics* 13, no. 1 (2020), 27–44.
Raschi, Francesco. "Prospettive euroscettiche: critiche all'Europa nella storia dell'integrazione," in *Prospettive euroscettiche: critiche all'Europa nella storia dell'integrazione* (Editoriale Scientifica, 2020).
Raschi, Francesco, and Elisa Piras. "A Realistic Scepticism: Raymond Aron's Perspective on the European Construction," *Journal of European Integration History* 26, no. 2 (2020), 267–284.
Schmitt, Olivier (ed.). *Raymond Aron and International Relations* (Routledge, 2018).
Stewart, Ian. *Raymond Aron and Liberal Thought in the Twentieth Century* (Cambridge University Press, 2020).

3 Karl Jaspers

Between the Bomb and Totalitarianism

Giunia Gatta

The timing of Karl Jaspers' reflections on the nuclear age is interesting. We find no mention of the atomic bomb in his works when the bomb was first dropped on Hiroshima and Nagasaki in August 1945. Even when he delivered his inaugural lecture to the University of Basel in 1948 on "Philosophy and Science," he did not refer to the new weapon, to its potential to destroy the world, or to the kind of philosophical reflections this new reality could spawn.[1] This was not unusual, especially in Europe around the end of World War II: relief for the end of the war trumped concerns about what a new war, fought with nuclear weapons, could bring.[2] Moreover, the creation of the United Nations as an institution aimed at containing war contributed to assuaging the fear that another use of nuclear weapons could be imminent. Certainly, some activism was underway from the beginning, especially among scientists who had developed nuclear technology,[3] and there is no reason to believe that Jaspers would have been unaware of it. But information about the consequences of the nuclear bombings of Hiroshima and Nagasaki was controlled, and pictures of the disfigured and the dead had been censored: the landscape of destroyed buildings in Hiroshima did not look too different to Britons and Germans from those of buildings in London or Berlin.[4] Moreover, Germany was too caught in the misery of military defeat and scarcity of everything in the wake of the end of the war to engage in debates about the bomb.[5]

The full appreciation for the potential of nuclear warfare occurred, then, not after the dropping of the first nuclear bombs, but over the 1950s, when the Soviet Union and the United Kingdom also acquired nuclear capability, and the United States tested the first hydrogen bomb. It is at this time that Karl Jaspers first intervenes on the question of the bomb in a speech broadcast on Basel radio in October 1956, entitled *Die Atombombe und die Zukunft des Menschen*, published two years later in 1958 in a much broader edition, also addressing the many questions listeners had asked him.[6] Since the beginning, Jaspers' concern for the atom bomb is entangled with a concern for "totalitarianism" in the Soviet Union. For him, the latter appeared as the biggest threat, to the point where he regarded pacifism as a problematic response to the nuclear threat. Pacifism probably seemed too close to appeasement, the great culprit for Hitler's rise to power and his unleashing of war upon the world, and the memory of how "totalitarianism" had ravaged Jaspers' life was too close in time, and Soviet hegemony too close in space for him to yield to the idea that disarmament was the wisest course of action.

Jaspers was a plausible pallbearer for the cause of "the West." In his first foray into political questions, written in 1930, he had expressed a late-Romantic support for individuality against the rising "masses" and a natural hostility towards communist ideals: "general equality" he regarded as "fiction" and constraining of greatness:

> Even an articulated mass always tends to become unspiritual and inhuman. It is life without existence, superstition without faith. It may stamp all flat; it is disinclined to tolerate independence and greatness, but prone to constrain people to become as automatic as ants.[7]

In the immediate aftermath of the end of World War II, in September 1946, Jaspers was invited – together with prominent intellectuals like Raymond Aron, György Lukáks, Maurice Merleau-Ponty, Stephen Spender, Jean Starobinski, and Jean Wahl – to the first edition of the Rencontres in Geneva, organized by intellectuals from the Swiss city to discuss the future of Europe. The *Rencontres* solidified Jaspers' stance as a committed defender of the West. By several accounts, the most salient exchange at the meeting was between Jaspers and Lukáks, an exchange which came to represent the nascent dialectic between East and West.[8] Lukáks had referred repeatedly to the 1941 alliance between the Soviet Union and Western democracies against fascism as a model, and he wanted to renew this alliance between "formal and real democracy," in order to win peace, just like war had been won. He attacked existentialism as the superstructure of the individualism typical of bourgeois society in its atomizing and privatizing drive.[9] Jaspers, on the other hand, was skeptical of socialism and instead envisioned a new world order in which Europe would maintain neutrality. In *The European Spirit* (1946), comprising his intervention in Geneva, he professes a lack of interest in the big political questions of the day, focusing instead on the individual dimension. Politics needs to purify itself and preserve as its only faith a faith in communication with the goal of the truth.[10]

This chapter offers an overview of Jaspers' thinking about the bomb and the entanglement of this theme with that of totalitarianism. Jaspers' approach to the danger of a nuclear holocaust consisted mostly of the superimposition of the philosophical framework he had developed over 40 years upon the new challenge posed by the bomb. But this superimposition found an important personal and political stumbling block: his assimilation of Soviet communism to Nazism – which had deeply affected him and his family – under the label of totalitarianism. Although his philosophical analysis about facing up to the bomb gives an important role to reflection and reason, we find in *The Future of Mankind* much unthinking embracing of the West and a dismissal of the Soviets as not even worthy of communication. This results in some internal contradictions that are not really solvable on philosophical grounds but only explicable on personal and political ones.

My contention is that Jaspers makes space, in the disjuncture between politics and what he calls suprapolitics, for ideological allegiances, which make his political analyses confusing because they are often inconsistent with his own philosophical premises. Like many philosophers, to borrow from Hannah Arendt, he helps us to think about what we are doing,[11] more than he helps us figure out what we should do. He is our guide into enduring the tension of insolubility more than in the quest for solutions.[12] There is certainly more value in this philosophical practice than meets the eye of the eager policy-maker, and some of his observations are remarkably timely: thinking before taking the plunge is important, and also insightful is the idea that political action ultimately does remain a plunge, impossible to parse out fully, least of all by the philosopher. But that is precisely why his unquestioned loyalty to "the West" can be puzzling for the reader.

Jaspers is more helpful *before* he articulates his policy recommendations. In his philosophical works in the first part of the 20th century, he notes that as human beings, we are all caught in situations that encompass us without us being able to fully comprehend them: death, suffering, guilt, struggle. We can evade them, but if we don't, they present us with the opportunity to grow more aware, and possibly more communicative human beings capable of solidarity.[13] There is something politically powerful in these ostensibly non-political insights, and something that can easily apply to pressing political problems such as an impending nuclear Holocaust, in the 1950s as today. But Jaspers appears to rush this mostly individual path to implausible scales. In *The Question of German Guilt*[14] and *The Future of Mankind*, he regards both the situation of Germany in the aftermath of World War II and the situation of the world in the shadow of the atomic bomb as boundary situations such as he had described in *Philosophy vol. 2*, as opportunities that call on a global (or national, in the case of Germany) scale for that existential elucidation which had been – in the philosophical writings – the prerogative of the individual. He was bitterly disappointed with the substantial failure of his fellow German citizens to take this path in coming to terms with their past, and this disappointment had a crucial role in his decision to move to Basel and become a Swiss citizen. He would most likely be bitterly disappointed, also in seeing that the danger posed by the bomb has itself become a missed opportunity for elucidation, and that many of the things he had advocated for in *The Future of Mankind* have not come to fruition.

Jaspers illuminates for us the profound truth that politics is meaningfully animated by our own aspirations and moral postures, and that this dimension significantly complements its more institutional and formal aspects. This is his biggest achievement as a political theorist. But this truth is then rushed to the political realm by way of a maladroit demand that everyone change their ways (hence an often-unsufferable prophetic tone), and it is articulated as necessarily leading to specific ideological stances. When Jaspers ventures into the terrain of policy, the result is often unclear or contradictory. I suggest, then, that we can think politically within his philosophical coordinates, but without being held hostage to the specific positions and expectations he articulates in his political works.

The chapter starts with an account of Jaspers' difficult relationship with politics in order to make sense of the ultimately unsuccessful blending of politics and philosophy in his thought, and then dives into Jaspers' major intervention in the question of the atom bomb, *The Future of Mankind*, addressing its uneasy mix of realism and idealism, the relationship between the nuclear and the totalitarian threat, and its juxtaposing of the call of reason and that of ideology. It concludes with a reflection on an alternative framework for addressing the threat of the bomb that one could draw from Jaspers' philosophy.

Jaspers' Uneasy Relationship with Politics

For the first 50 years of his life, Karl Jaspers could afford to not give politics much thought. He came from a relatively privileged family, and due to a chronic pulmonary condition, he was spared from fighting in World War I (something he felt guilty about). He regarded political entanglements with disdain and even reproached his friend and mentor, Max Weber, for "wasting his time" in such matters.[15] This sense of superiority towards politics came to a halt with the rise to power of Adolf Hitler, when politics repaid Jaspers' disdain by threatening his intellectual life and even his own survival and the survival of the people he loved. Jaspers was married to a Jewish woman to whom he was extremely devoted, so during those years he was stripped of his professorship and prevented from teaching and publishing anything. He also consistently feared for his and his wife's lives. The sense of powerlessness in the face of politics that he felt during those years persisted even when the Nazis were defeated and he took on an important role in the reopening of German universities, and Heidelberg in particular.[16] Politics remained for him the somewhat overwhelming realm of force. He consistently believed that the only hope for freedom to burst in this realm was by way of internal changes within each individual, and even that burst could not but be episodic. Certainly, he thought that there were no indigenous resources within the political realm that could bring about change in the right direction.[17]

Jaspers had already articulated the belief, in the early 1930s, that only a spiritual and cultural revolution could solve the predicament in which the modern world found itself.[18] This stance characterized his approach to politics for the rest of his life.[19] Jaspers' precarious state of health throughout his life, and the tumultuous times in which he lived, made him very aware of the ever-present possibility of catastrophe. While he did not foresee a state of the human condition where this possibility would be overcome, he did have faith in the capacity of individual human beings to elucidate their existence to the point where they would be aware of their own finitude and the finitude of their fellow human beings. This awareness, to be gained especially through existential communication with others, could bring about solidarity and change, if a preponderant number of people were to achieve it.[20] Jaspers seemed less interested, even in his more political writings, in the more properly "political" and operational mechanisms of change: his focus was rather on its "suprapolitical" basis.

The political domain is, for Jaspers, the domain of the calculating intellect, of regularity, and of force. He was impatient with an approach to politics as the "slow boring of hard boards,"[21] and consistently looked for that spiritual revolution that would somehow fix things. But this solution had origins different from those internal to politics. In this sense, he also differed – in his understanding of politics – from his student and friend, Hannah Arendt. For her, politics is the realm of appearances and freedom that suffices to itself,[22] whereas for Jaspers, it is the realm of force dependent on suprapolitics for any inspiration. If, for Arendt, action in concert is the highest incarnation of freedom,[23] and proof of politics as an autonomous realm, for Jaspers, ordinary political action is constrained by force and responds to the ineluctable regularities of the intellect. The intellect, unable to grasp inner change, cannot understand the only real opportunity for redemption available to humankind in the atomic age. For these kinds of insights, humanity needs to rather turn to reason as a more comprehensive and existential faculty.[24]

Although much of *The Future of Mankind*, the text in which Jaspers most directly deals with impending nuclear catastrophe, depends on the relationship between politics and suprapolitics, the path between the two is unclear, and Jaspers does not dwell on their connections. He rather employs metaphors of sudden and somewhat arbitrary transition – remarkably different from the Weberian "slow boring" – and without really indulging in discussions about natality, the term that Arendt embraces to describe the burst of political action into the world.[25] Political action is instead for Jaspers a "plunge into the tide of history": far from reassuring, and ultimately "undeducible."[26]

In *The Future of Mankind*, Jaspers feels dragged into a political realm he felt foreign to, for the second time in the span of just over ten years.[27] The first time, with *The Question of German Guilt*, the foray into the political realm cost him the sense of feeling at home in his country, due to the cold reception of his fellow Germans to his call to admit to their guilt and change. This second foray was prompted by a convergence of political circumstances in the shadow of the Soviet Union's acquisition of nuclear power: most prominent among them were the Soviet invasion of Hungary and the Suez crisis, both of which Jaspers refers to repeatedly in the book. Jaspers supported the rebellion in Hungary and Israel's invasion of Gaza and the Sinai peninsula, but he admitted in correspondence with Arendt to being a "dilettante," whereas he regarded both Arendt and her husband Heinrich Blücher as "experts." He regarded his "atom bomb essay" as an "expression of political consciousness in the shadow of the H bomb," but he was not altogether satisfied with it.[28] He was therefore eager to have Arendt's response to the book, where he had "taken a chance with a subject for which the extent of my knowledge is inadequate."[29]

Let us now turn to that book for a closer reading.

The Future for Mankind, between Hard Realities and Lofty Aspirations

There is something paradoxical about a philosopher who has maintained throughout his life that struggle is an ineliminable feature of the human condition,[30] and the conviction of the same philosopher that the only way to escape the annihilation of humankind by nuclear bombs is the abolition of war. *The Future of Mankind* is not a pacifist book. It is a book in which both elements of the paradox remain in place, and where the call to forceful military resistance frequently overshadows the need to abolish war. This latter goal must be deferred, for Jaspers, until the survival of "the West" is secured, and the abolition of war is to be obtained not by way of diplomatic means, or direct pressures on powerful politicians, or the creation of a world government (everything Bertrand Russell spent his life advocating for), but rather through the progressive enlightenment of individuals all over the world, who must all undergo a form of inner change that will effectively make war impossible. In the meantime, suggests Jaspers, let's fight for the survival of the West.

Judith Shklar's review of Karl Jaspers' *The Future of Mankind* is scathing. It is, she writes, standard Cold War ideology in dramatic tones. The quick reviews of anything from the United Nations, to Gandhi, to colonialism, do not rise – according to her – above the level of the Sunday newspaper supplements.[31] Shklar is right about many things in her review: the tone Jaspers uses is often dramatic. At times, preachy. And often his insights on the political situation of the time do not reach particular depth or run into contradictions perplexing for the reader. Shklar is also right when she notes that we find in *The Future of Mankind* an application of Jaspers' earlier philosophy, which she finds in perfect syntony with the compulsive defenses of "the West" during the Cold War. As mentioned earlier, Jaspers certainly had an "all or none" posture towards politics, as either purely the realm of force or as the ground for the flourishing of what he called "suprapolitical" principles, drawn from ethics and a spirit of sacrifice.

The Future of Mankind moves on two registers. On one hand, Jaspers advocates for political instruments to cope with the appearance of the nuclear bomb in the world: he calls for diplomacy, for the establishment of agencies to control the use and development of the new weapon, and for devising institutional techniques to make treaties effective. But most fundamentally, he regards those arrangements as not enough,[32] believing that something *above* politics is needed to make politics work and bring human beings away from the brink of total destruction. This is "the ethical idea" and "the valor of self-sacrifice," which are, for Jaspers, "suprapolitical" motives inspired by what Jaspers calls "reason."

Jaspers tells us that we must radically change ourselves, but in the face of this radical demand, we are not really told how this will bring about change at the global level (other than awareness of the brink of disaster) and lead to the abolition of war. The transition from extreme realism to what we might legitimately call extreme utopianism is left to individuals and their inner change. But by Jaspers' own admission, "the moral idea cannot be planned," and "change of heart cannot be forced," which is also why we are not really given specific pathways to our own change or that of humanity in general.

The Future of Mankind is divided into three parts. The first part was, in Jaspers' mind, a reconstruction of discussions that – in his opinion – lead to dead ends. In the second, he reviews the present world situation, and in the third, he sets out to "clarify human existence." In practice, similar ideas crop up in all three parts, most notably the idea that "suprapolitics" needs to take the helm in order to guide politics towards less destructive ways. Technology is regarded throughout with considerable suspicion, and as something to be governed lest it destroys us.[33] Moreover, and understandably given this individual focus, a lot appears to hinge on the qualities of individual statesmen.

The problem with political controls on the use of the bomb, Jaspers writes in the first part, is that they imply the solidarity they would create.[34] Jaspers tackles here a fundamental circularity in political philosophy and politics in general: human beings need rules to govern their lives, but in order to come up with those rules and obey them, they need to somehow already be the kind of human beings that those rules are trying to constitute. As he puts it later on in the book: "the trouble is that democracy, which is to make the people rational, presupposes a rational people."[35] His take on breaking this circularity is for each human being to work on themselves so that they will embrace sensible rules, and the rules will thereby be effective.

The basic starting point of this process of renewal for each human being is sensitivity to wrong and injustice, and the mustering of the energy to correct them. Yet, this sensitivity and energy will rise in a context that is inevitably compromised by injustice, so righteousness cannot suffice for survival: there is always a remnant of force that will need to be employed.[36] The general goal is to keep the wrong from attaining proportions that make violent rebellion irresistible, and the perspective here cannot be that of single nations or peoples, but that of the entire world.[37] In any case, for Jaspers, mere politics will not get us to this sensitivity. We need what he calls the ethical idea and the value of self-sacrifice.[38]

The ethical idea and the value of self-sacrifice are regarded by Jaspers as interconnected modalities of relationships with others that would make war impossible.[39] They escape, in Jaspers' view, the logic of success and failure that reigns in the political realm, so while they are not typical political instruments, they have decisive political consequences.[40] The ethical idea and the idea of self-sacrifice in themselves lead to embracing nonviolence as a value. But in the context of the Cold War, contemporary to the drafting of the book, Jaspers believed they could not come to fruition, so his stance on nonviolence is surprisingly ambivalent.

The translation of nonviolence in the political realm presupposes the confidence that readiness to suffer violence will conquer violence. When practiced in the context of India's struggle for independence, two fundamental elements brought it to success. First, Gandhi rooted the practice of nonviolence in suprapolitics: he faced force and was able to challenge it by convincing the Indian masses to suffer the consequences of force and to renounce violence.[41] Second, and crucially for Jaspers, Gandhi's undertaking occurred in the context of the British Empire, whose liberal views and legal concepts "gave Gandhi his scope."[42] Jaspers goes as far as to attribute the liberation of India more to benevolent logics within the British Empire than to Indian agency and Gandhi's agency specifically. This brings him to discount the extent to which nonviolence could be an instrument to deal with the rise of the atomic age, in the context of the Cold War and Soviet totalitarianism.

In a totalitarian context, such as the one in Germany and in the Soviet Union, the kind of sacrifice demanded by nonviolent practices loses political traction because there is no publicity for it, so it remains irrelevant.[43] On this basis, Jaspers berates the pacifists of his time, for he believes that in the face of force, especially the absolute force of totalitarianism, "nothing remains but force or submission to force."[44] Politics at the time of his writing presents itself to him as a "zero sum game," where any concession entails submission. Passivity is not a sensible approach, nor is disarming while others arm.[45] Abolition of war is the goal if we are to save ourselves from complete destruction, but "totalitarianism" intervenes to essentially take the possibility off the table. This is why, by the end of the first part of the book, totalitarianism enters prominently into the discussion to intersect, but also in a way displace, the question of the bomb: any response to the predicament of the atomic age needs to navigate the strait between nuclear holocaust and what Jaspers construes as loss of freedom.

Totalitarianism and the Bomb

Totalitarianism and the bomb were born as alternatives. After all, the bomb was developed in the United States by scientists who were moved by the need to defeat Nazism, many of them refugees from Europe. For Jaspers, the bomb has not yet exhausted its work as a bulwark against totalitarian regimes.

Despotism stands in the way of rational man, as Jaspers conceived him, and today it specifically stands in the path to the rescue of mankind. We cannot give up the bomb, lest we give in to totalitarianism. Caught in this strait, citizens of the "free world" ought to shake off the yoke of a "life lost in an empty consumer's existence after a few hours of unrelished work"[46] and turn to a more authentic way of living so that, with their authenticity, they will win over those subjected to totalitarian rule. The problem, as Jaspers sees it, is that the clash between the two superpowers is a clash between truth and falsehood, but "while totalitarianism rests upon the principle of falsehoods and thus grows mighty, the free world is not truthful enough and thus grows weak."[47]

Jaspers takes the coordinates for approaching totalitarianism from Arendt's *The Origins of Totalitarianism*, but much more than in Arendt's work this notion turns into Cold War fodder.[48] Jaspers underscores the centrality of lying to this modality of government, against which he envisions "the spiritual power" of freedom as the key to victory, not only against the Soviet Union but also against "subversive organizations" within the free states.[49] Once again, this battle depends not on methods of repression that would turn the free world into totalitarian itself, but on the cultivation of individual integrity:

> Every individual in the free world who appears unconvincing to the other, who displays the seamy side of misused freedom in lawlessness, arrogance, and greed, inflicts a defeat upon freedom by his very existence. The free world can be saved only if its members prove its value.[50]

We can see here a drive towards the moralization of politics and indeed of every single individual, who "by his very existence" may threaten the existence of the "free world." There is no reflection here on exploitation and structural injustice, which may be what brings "malcontents, rebels, and desperadoes in the world" to hatred and destruction.[51] Jaspers reframes the opposition between capitalism and Marxism into a contrast between totalitarianism and freedom. There is also, as Shklar had lamented in her review, an undiscriminating assimilation of communism and Nazism under the sweep of totalitarianism, and no serious consideration of the plea of those whom "freedom" may leave behind. Or, for that matter, consideration for anyone who does not align to the idea of freedom put forth by America, "even when she is wrong."[52]

"We" is for Jaspers "the West," which indeed he construes bowing to traditional Cold War ideology. Who is *in* the West? Who is *out*? Generally, according to him, anything good we may find even in non-traditionally Western countries is due to Western influence. Russia is not hopelessly lost to totalitarianism, for it broadly speaking "belongs to the West" and the Russians are "white people." Yet, Russia is Western and Asian at the same time, and it "never knew the political liberty of the West" although "under

Western influence has produced great writers of her own who belong to world literature."[53] But as the mention of whiteness denotes, the constitution of the willing around freedom, as the argument progresses, takes the form more of *ethnos* than culture.

Former colonies are caught in the middle between the West and totalitarianism, for Jaspers. His treatment of colonialism is emblematic of his uneasy navigation between philosophy, politics, and ideology. He seems to indict the legacy of colonialism and discusses the opportunities and dangers offered by its end. But the extent of interest for former colonies soon reveals itself as being an interest in enlisting them as pawns in a geopolitical competition between the two world powers. Jaspers, for example, understands the fear of "the West" that former colonies, "still largely incapable of freedom,"[54] will fall prey to totalitarian Russia. In order to contrast this process, "the West" will have to sincerely give up any colonial motivation and change, leading by example but also standing united under the aegis of America, again, "even if she is wrong"[55] for he reminds his readers that each Westerner has his own country, and the country that is the sure foundation of their political reality, the United States of America.[56] Jaspers does not see any contradiction in discussing the need for decolonization together with the proclamation that absolute loyalty is due to one of the two superpowers, even though at the time of his writings the United States had already engaged in problematic foreign policy operations, with a strong neocolonial flavor, especially in Latin America. Then again, Jaspers sees nothing wrong (and rather advocates for) "Western settlements" in the Sahara to protect oil, in the name of Western solidarity, and regards those settlements as akin to those of "the pioneers that built America."[57]

Jaspers does believe that injustice threatens world order, for in his opinion, injustice breeds recourse to force, since the injured regard the legal status quo as "nothing but a form of fatal, lawless force."[58] This attempt at rebalancing world order is what causes war and what the United Nations, in Jaspers' opinion, has not been able to address. He describes the United Nations as a toothless tool in the hands of those who actually do not want to take responsibility for redressing the wrongs of the world. It appears, however, that in Jaspers' mind, settlements to protect Western interests in non-Western countries do not constitute an injustice. In fact, he criticizes what he calls the legalistic concept of "aggression" as the only ground for which the United Nations allows the use of force. He notes how violations of foreign rights within territories recognized as sovereign (like those Western settlements in the Sahara, we can presume, or British control of the Suez Canal) affect world peace as well and should also be considered aggression. He also wants to expand the concept of aggression to include arming at the border and declaring no right to the existence of another state.[59] Use of force in those cases is warranted, for if force were not used "we would breed another, threatening force, protecting its borders until it was ready to cross them. Is it aggression or self-defence if a small state breaks out of its encirclement at the last moment?" Jaspers here has Israel in mind.

Jaspers often mentions that any solution to humanity's current predicament will have to be grounded in the world as it now is, which is why he rejects the idea of a world government, a dangerous idea also because it would have to be based on phenomenal despotism and conquest in order to survive. Force must remain plural and decentralized if it is to be constrained by law.[60] Instead of striving to constitute a world government, we ought to aim at a confederation of nations living under free constitutions. At the moment he is writing, Jaspers believes – however – that there cannot be a world confederation, but only a hope to attract countries that are not yet under the sway of "totalitarian" Russia.

What Is to Be Done?

In the last part of *The Future of Mankind*, Jaspers addresses possibilities for action given the current situation of the world. He challenges the view that the answer to humanity's current predicament and danger of annihilation by the bomb can be provided by science. Science is a matter of expertise and study acquired over a long period of time, and as such it is not of pertinence to the whole public, but Jaspers sees the solution to our current predicament as being of general concern, and not just the concern of scientists. Moreover, and more essentially, science is the domain of the intellect, which trades in regularities.[61] Salvation, however, cannot come from regularities, but rather from a deviation from regularity. We need, instead, a philosophical disposition – which for Jaspers is the responsibility of everyone – to illuminate what we already know and deviate from the intellect to reason so that we can existentially transform ourselves.[62] This is not, Jaspers reminds us, a matter of studying philosophy, but of actually philosophizing, without an immediate concern for the utility of the process.[63] While the intellect walks on preordained causal tracks, reason accommodates freedom and new thinking: "to turn back from the accustomed, self-sufficient intellectual way of thinking to the way of reason – that is the change in man on which his future depends."[64]

Rational thinking exists, for Jaspers, only in common: the individual cannot be rational by themselves.[65] In fact, Jaspers envisions the creation of a community of men of reason, who will animate institutions that otherwise would remain nothing but empty shells. Freedom "speaks in communication."[66] Jaspers is not here talking about political, but of existential freedom; not about political, but existential communication. These understandings of existential freedom and communication are prepolitical and suprapolitical, and yet animate all forms of dignified politics: "the non-contractual but actual community of reason needs to prevail in all forms of human order – in states, parties, churches, schools, unions, and bureaucracies. It cuts across all conflicts, whether religious, partisan, or international."[67] Where the intellect has brought us to the brink of disaster, reason can hold us back from it. It does so not by telling us exactly what to do, but by turning us into reflective beings who meet others as peers and as potentially equally rational beings.[68]

Jaspers addresses the objection that a change in man does not amount to a change in politics at large by simply restating that change, if it will come, will not come from "objective sociopolitical processes,"[69] but precisely from change within individuals, and in particular individuals in positions of leadership. Change will affect first "a few, then many, and finally, perhaps, a majority."[70] Drawing a distinction between the politician and the statesman, he attributes to the latter all the characteristics that might avert nuclear doom. He mentions as examples Pericles and Moses and appears to reject some of the attributes that Machiavelli extols in his model Prince. The statesman looks instead like the leader Weber discusses as a politician by vocation and also bears some resemblance to Rousseau's Great Legislator. He serves as an example and as a source of inspiration for the citizens, "implanting" in their hearts the crucial issues. He represents the ethos of the people and is able to awaken in them those moral-political principles that "lie hidden in them, but ready."[71] He does not take public opinion as a given but rather *creates* it through an education to reason. Yet, he must not be deified lest the purpose of his action – the expansion of freedom – be undermined. The statesman does not let ideology limit his horizon and is aware of the value of clashing philosophical thoughts. For him to succeed, his whole life is open to scrutiny and must be impeccable, publicly and privately.

The rejection of contractualism and the view of a community of reason that will begin with the enlightenment of a few, who will by example – but also perhaps by manipulation – create an ethical community in their own image, reveal a suspicion for the masses with Platonic and Burkean inflections. Reason, while potentially inherent in any human being, is not likely to really take hold of humanity as a whole (at most, it seems that we can hope for a majority). In Jaspers' repeated proclamations, reason stands for freedom, open communication, leadership, and radical change within individuals, and it has a special relationship with democracy, which Jaspers defines as "the system in which the people – and peoples – are supposed to 'reason together.'"[72] Sure: reason, Jaspers writes, can prevail reliably only "if it guides the people along with their leaders."[73] Democracy's ability to put itself into question and to improve shows its link to reason: it is animated by sensitivity to injustice, and "lives by the active solicitude that makes a wrong done to one a matter of concern to all."[74] But referring to de Tocqueville's and Weber's critique of some aspect of democracy, Jaspers notes that "no democratic form of government can guarantee the democratic idea."[75] We can never be sure that democratic institutions will uphold democratic principles: "the same institutions can be used to save democracy and to destroy it," so we cannot rely on legal guarantees alone.[76]

Everyone is responsible for their government's actions and for their own self-education. In a democracy, I cannot shift responsibility onto the majority for decisions I do not embrace. As he expressed in *The Question of German Guilt*, Jaspers believes that – especially in democracies – all citizens are politically liable for the consequences of democratic decisions, even if they may not be all equally guilty. They are liable because they are responsible for bringing their best to power. Jaspers is skeptical of radical democracy as a way to achieve this goal, even as he embraces democracy overall. Democracy must retain some hierarchy if it is to survive. While the democratic ethos demands that no one be despised or idealized, there is no real equality of talent.[77] Democracies should oppose unjust privilege but "foster their own aristocracy."[78] And indeed, Jaspers believes that the strength of freedom rests "on democratic recognition of human rank,"[79] so he advocates for separate schools or classes for the gifted, as a matter of justice towards them but also really to everyone.

For all its problems, democracy thus understood remains the best chance for a group of human beings to grow rationally together, even as it may entail a leveling that risks its perversion "into one of the worst dictatorships ever known."[80] In the end, however, we must remember that the goal is human freedom, and democracy in the sense in which Jaspers clarifies it is the best tool to achieve it.

Reason and Ideology

Reason keeps the degeneration of democracy in check because it stands against ideology and intolerance by providing ground and the conditions for communication. Yet Jaspers cannot see how Cold War ideology seeps into his portrayal of what is rational in the realm of politics and does not recognize as ideology his own political stance. On p. 277 of *The Future of Mankind*, for example, he suggests that it is reason that makes us see "the important distinction between indispensable special planning for the moment and fatal total planning for an unattainable entirety." Reason appears to have substantive content, a content that overlaps broadly with the ideology of "the West." He also does not see as ideological his dismissal of Marxism as scientific superstition and as the foundation of totalitarianism in his time, oddly blaming it for its determination to change man and its prophetic tone and despotic way of thinking, some of which Jaspers seems to be advocating.[81] Marxism in Western countries is for Jaspers not even part of the ideological struggle, and a useful counterpoint and challenge to the mainstream, but "stealthy poison."[82] One wonders whether the openness Jaspers extols in reason is compatible with the rigid binary that he poses between freedom and totalitarianism. It is easy, given Jaspers' life story, to understand his unquestioning allegiance to the United States, and his indictment of any form of appeasement as "totalitarianism." More difficult is it to fit both within his philosophical coordinates, premised on reflectiveness, self-criticism, and suspicion for absolute truths. His seemingly unquestioning embrace of one side in the Cold War struggle is all the more puzzling when he suggests that reason "demands modesty," and that "a large part of reason is not to mistake our own thoughts and wishes for reason itself."[83]

How, then, can we reason today with Jaspers about a looming nuclear catastrophe? His best political insights, I believe, are drawn from his strictly philosophical works. He had identified "boundary situations" as those definitive traits of being in the world that shock us out of the unreflective routine of our lives and bring us to see ourselves both in the concreteness of our embodied existence and in the universality of our experience as human beings. In *Psychologie der Weltanschauungen* (1919) and in *Philosophy vol. 2* (1932), he specifically discussed the fact of always being in a situation, suffering, death, guilt, and struggle as such very concrete, but also universal situations. Clearly, Jaspers saw the dawn of the atomic age as itself a boundary situation, but caught in a somewhat bland and repetitive call for man to change and in the ideological struggles of his day, he failed to either pursue the path of elucidation, communication, and solidarity that the bomb could bring about, or explore the entanglement of the bomb with the other boundary situations he had presented in his more strictly philosophical works: how can the very concrete and specific suffering and death of those who have already been victimized by the bomb bring us consciousness of the nuclear catastrophe so that we might garner the political will to avert a new use of the bomb? What are the implications of an assumption of guilt by some of the most powerful countries and citizenries of the world with respect not only to the deployment of the bomb but also to the enduring injustices of the world, be they global in terms of the legacy of colonialism, or local in the patterns of economic or racialized discrimination within each country? And finally, what can the ties between the inevitability of struggle and the asymmetries of power that characterize our world tell us about promising directions for change?

These themes do make some appearance in Jaspers' treatment of the atom bomb, but they are suffocated by Jaspers' somewhat contingent ideological commitments. We don't have to fall prey to the same commitments and reach his same conclusions but can rather use the philosophical instruments he provides to recover the ground for political communication opened by our proximity to the abyss, by our being at the boundary of catastrophe. Maurice Blanchot notes in a discussion of Jaspers' book that in the face of such a novel and radical challenge as the advent of a weapon that can destroy humanity and the planet, Jaspers returns to calls and conceptual frameworks he had formulated, identically, decades before.[84] The question, of course, is broader than Jaspers' book: why is it that we cannot face these deep existential questions with the gravity they deserve? The bomb, but today also the threat of environmental collapse? Blanchot notes that while these threats haunt us as a totality, as a possibility of destruction, humanity does not yet exist as a whole, so does not have power over those weapons it invented. Humanity can be wiped out, but not affirmed.[85] In its potential to awaken us to new ways of thinking, then, the "apocalypse" turns out to be "disappointing." Giorgio Agamben notes, discussing Blanchot's essay, how indeed the possibility of the end of humanity has become over time an obvious and almost trivial fact, scarcely commanding the intellectual energy of philosophers, let alone the public. The continuous invocation of emergency has transformed the exception into the rule, and our thought no longer measures up to the possibility of catastrophe.

I believe that Blanchot is right in his critique of Jaspers and in his observation that we have not yet developed a new way of thinking and a new way of speaking to make sense of the possibility that humanity might end. Yet I also believe that Jaspers' own philosophy, and his categories of boundary situation and communication, hold much promise in pushing us to ask the right questions and even come to some political answers. While he was right that a political route might be taken that would bypass his demand for a change in man and would simply focus on getting used to a state of undischarged high tension, he warned that fear and deterrence would not be enough, for nothing is impossible if the captains – against common sense, against reason, against the moral qualms that inhibit even criminals – decide to drag everyone down with them. When Hitler saw he was doomed, he plainly wanted to doom the German people, too, while he had the power. Collective suicide is not out of the question if leaders meet in loathing or indifference or blind destructiveness, or if but one of them feels that way. They may slide into the abyss as they slid into war in 1914.[86]

The difference with previous world wars, of course, is that the destruction would involve the entire planet, in 1958 as in 2025. Yet, and contrary to what he hoped for, while practices of inner change and renewal might help those who undergo them, and even have some exemplary value, there cannot be a certain path to security. If fertile thinking about the apocalypse is meant to secure us from it, then it will happen in vain. The willingness to engage our fellow human beings, and work politically with them to find ever-temporary solutions and re-compositions of conflicts, citizens and statespersons alike, is our only hope, and one that I find in keeping with the thrust of Jaspers' thought overall, and possibly the thought of Blanchot and Agamben as well. As intellectuals, one of our most important tasks is to undermine the fallacy of perfect security, for all wars up to the most recent ones flourish on this ultimately unattainable wish.

Notes

1 Karl Jaspers, "Philosophy and Science," in *Way to Wisdom*, ed. Karl Jaspers (New Haven, Yale University Press, 1960), 147–167.
2 Holger Nehring, *Politics of Security: British and West German Protest Movements and the Early Cold War, 1945–1970* (Oxford, Oxford University Press, 2013), 16.
3 Dexter Masters and Katharine Way, eds., *One World or None: A Report to the Public on the Full Meaning of the Atom Bomb* (New York, New Press, 2007, 1946).

4 Nehring, *Politics of Security*, 20.
5 The correspondence between Jaspers and Hannah Arendt, his former student, testifies to this paramount concern for survival, the focus being primarily on the care packages that Arendt was regularly sending to Jaspers and his wife in the immediate aftermath of the war's end. See Hannah Arendt and Karl Jaspers, *Correspondence 1926–1969* (New York, Harcourt, 1992).
6 Karl Jaspers, *Die Atombombe und die Zukunft des Menschen* (München, Piper, 1958), translated as *The Future of Mankind* (Chicago, The University of Chicago Press, 1961).
7 Karl Jaspers, *Man in the Modern Age* (New York, Anchor Books, 1957), 40.
8 Gianfranco Contini, *Dove va la cultura europea?* (Macerata, Quodlibet, 2012); Stephen Spender, "The Intellectuals and the Future of Europe," *The Gate/Das Tor* 1 (January/March 1947), 2–9; Suzanne Kirkbright, *Karl Jaspers. Navigations in Truth. A Biography* (New Haven, Yale University Press, 2004), 208–214; Anna Pia Ruoppo, *Marxismo ed esistenzialismo: due filosofie dell'Europa. Lukács e Jaspers si incontrano a Ginevra* (Milano, Mimesis, 2023).
9 Contini, *Dove va la cultura europea?*
10 Karl Jaspers, *The European Spirit* (London, The Stanhope Press, 1948).
11 Hannah Arendt, *The Human Condition* (Chicago, The University of Chicago Press, 1958), 5.
12 Jaspers, *The Future of Mankind*, 12.
13 Karl Jaspers, *Philosophy vol. 2* (Chicago, The University of Chicago Press, 1970)
14 Karl Jaspers, *The Question of German Guilt* (New York, Fordham University Press, 1965).
15 Karl Jaspers, *On Max Weber*, ed. John Dreijmanis (New York: Paragon House, 1989), xv. I have discussed Jaspers' political stances more in depth in Giunia Gatta, "Between Politics and Suprapolitics: Karl Jaspers and the Flight from Force," *Political Science Reviewer* 42, no. 1 (2018), 119–134.
16 Mark W. Clark, "A Prophet without Honour: Karl Jaspers in Germany, 1945–48," *Journal of Contemporary History* 37, no. 2 (2002), 197–222.
17 Gatta, "Between Politics and Suprapolitics," 124.
18 Karl Jaspers, *Man in the Modern Age*.
19 Christopher Thornhill suggests that *Man in the Modern Age* would become over time a source of embarrassment for Jaspers, but although he contextualizes the book in his time, in a Preface to a new edition from 1951 Jaspers maintains that the book "was as valid now as then, in spite of all that has happened since its first appearance." See Chris Thornhill and Ronny Miron, "Karl Jaspers," in *The Stanford Encyclopedia of Philosophy*, eds. Edward N. Zalta and Uri Nodelman (Summer 2024 Edition). Available at https://plato.stanford.edu/archives/sum2024/entries/jaspers/ and Karl Jaspers, *Man in the Modern Age*.
20 For the most extensive treatment of Jaspers' idea of existential elucidation, see his *Philosophy, vol. 2*
21 Max Weber, "Politics as a Vocation," in *From Max Weber: Essays in Sociology*, ed. by H.H. Gerth and C. Wright Mills (New York, Oxford University Press, 1958), 128.
22 Arendt, *The Human Condition*, 198–199.
23 Arendt, *The Human Condition*, 179.
24 Jaspers, *The Future of Mankind*.
25 Arendt, *The Human Condition*, 8.
26 Jaspers, *The Future of Mankind*, 11.
27 Hannah Arendt and Karl Jaspers, *Correspondence 1926–1969* (New York, Hartcourt, 1992), 310.

28 Arendt and Jaspers, *Correspondence*, 308. Consistently, Jaspers, who had two brothers-in-law living in Israel, expresses in the correspondence an enthusiastic support for Israel, in contrast with Arendt's skepticism or outright condemnation (see especially letter 233 on p. 358).
29 Arendt and Jaspers, *Correspondence*, 350.
30 See especially Karl Jaspers, *Psychologie der Weltanschauungen* (Berlin, Springer, 1919) and Jaspers, *Philosophy vol. 2*.
31 Judith Shklar, review of *The Future of Mankind*, *Political Science Quarterly* 76, no. 3 (1961), 437.
32 Jaspers, *The Future of Mankind*, 22.
33 This is a theme that Jaspers had already addressed in *Man in the Modern Age*, as Mats Andrén notes in his "Karl Jaspers on the Atomic Bomb and Responsibility," *Existenz* 14, no. 2 (2019), 6.
34 Jaspers, *The Future of Mankind*, 17.
35 Jaspers, *The Future of Mankind*, 294.
36 Jaspers, *The Future of Mankind*, 18.
37 Jaspers, *The Future of Mankind*, 19.
38 Jaspers, *The Future of Mankind*, 23.
39 Jaspers, *The Future of Mankind*, 28.
40 Jaspers, *The Future of Mankind*, 30.
41 Jaspers, *The Future of Mankind*, 36–37.
42 Jaspers, *The Future of Mankind*, 38.
43 Jaspers, *The Future of Mankind*, 39.
44 Jaspers, *The Future of Mankind*, 40.
45 Jaspers, *The Future of Mankind*, 52.
46 Jaspers, *The Future of Mankind*, 56.
47 Jaspers, *The Future of Mankind*, 93.
48 Jaspers, *The Future of Mankind*, 104; Hannah Arendt, *The Origins of Totalitarianism* (New York, Harvest, 1976).
49 Jaspers, *The Future of Mankind*, 107–108.
50 Jaspers, *The Future of Mankind*, 110.
51 Jaspers, *The Future of Mankind*, 110.
52 Jaspers, *The Future of Mankind*, 92.
53 Jaspers, *The Future of Mankind*, 119.
54 Jaspers, *The Future of Mankind*, 86.
55 Jaspers, *The Future of Mankind*, 92.
56 Jaspers, *The Future of Mankind*, 101.
57 Jaspers, *The Future of Mankind*, 148.
58 Jaspers, *The Future of Mankind*, 148
59 Jaspers, *The Future of Mankind*, 150.
60 Jaspers, *The Future of Mankind*, 96–97.
61 Jaspers, *The Future of Mankind*, 196. In a letter from April 1957, Arendt had expressed gratitude for the Göttingen Declaration, signed by 18 German physicists against the possibility that Germany would acquire nuclear weapons, which also proclaimed the unwillingness of the scientists to cooperate with such a possible acquisition in any way. Jaspers replied in a somewhat petty way, expressing his lack of trust for "these people (based on personal experience)" and dismissing the appeal of the scientists as inconsequential. Arendt and Jaspers, *Correspondence*, 313 and 315.
62 Jaspers, *The Future of Mankind*, 204.
63 Jaspers, *The Future of Mankind*, 209.
64 Jaspers, *The Future of Mankind*, 217.
65 Jaspers, *The Future of Mankind*, 218.

66 Jaspers, *The Future of Mankind*, 214.
67 Jaspers, *The Future of Mankind*, 224.
68 Jaspers, *The Future of Mankind*, 260–261.
69 Jaspers, *The Future of Mankind*, 233.
70 Jaspers, *The Future of Mankind*, 261.
71 Jaspers, *The Future of Mankind*, 240.
72 Jaspers, *The Future of Mankind*, 291.
73 Jaspers, *The Future of Mankind*, 292.
74 Jaspers, *The Future of Mankind*, 293.
75 Jaspers, *The Future of Mankind*, 294.
76 Jaspers, *The Future of Mankind*, 295.
77 Jaspers, *The Future of Mankind*, 310.
78 Jaspers, *The Future of Mankind*, 311.
79 Jaspers, *The Future of Mankind*, 314.
80 Jaspers, *The Future of Mankind*, 299.
81 Jaspers, *The Future of Mankind*, 264 and 278.
82 Jaspers, *The Future of Mankind*, 278.
83 Jaspers, *The Future of Mankind*, 307.
84 Maurice Blanchot, "The Apocalypse Is Disappointing," in *Friendship* (Stanford, Stanford University Press, 1997), 101–108, here at 103. See also Giorgio Agamben, "La guerra atomica e la fine dell'umanità," *Quodlibet*, October 7, 2022. Available at www.quodlibet.it/giorgio-agamben-la-guerra-atomica-e-la-fine-dell-u2019umanita
85 Blanchot, "The Apocalypse is Disappointing," 106.
86 Jaspers, *The Future of Mankind*, 322.

Bibliography

Agamben, Giorgio. "La guerra atomica e la fine dell'umanità," *Quodlibet*, October 7, 2022. Available at www.quodlibet.it/giorgio-agamben-la-guerra-atomica-e-la-fine-dell-u2019umanita

Andrén, Mats. "Karl Jaspers on the Atomic Bomb and Responsibility," *Existenz* 14, no. 2 (2019), 1–9.

Arendt, Hannah. *The Human Condition* (Chicago, The University of Chicago Press, 1958).

Arendt, Hannah. *The Origins of Totalitarianism* (New York, Harvest, 1976).

Arendt, Hannah, and Karl Jaspers. *Correspondence 1926–1969* (New York, Hartcourt, 1992).

Blanchot, Maurice. "The Apocalypse Is Disappointing," in *Friendship* (Stanford, Stanford University Press, 1997), 101–108.

Clark, Mark W. "A Prophet without Honour: Karl Jaspers in Germany, 1945–48," *Journal of Contemporary History* 37, no. 2 (2002), 197–222.

Contini, Gianfranco, *Dove va la cultura europea?* (Macerata, Quodlibet, 2012).

Gatta, Giunia. "Between Politics and Suprapolitics: Karl Jaspers and the Flight from Force," *Political Science Reviewer* 42, no. 1 (2018), 119–145.

Jaspers, Karl. *Psychologie der Weltanschauungen* (Berlin, Springer, 1919).

Jaspers, Karl. *The European Spirit* (London, The Stanhope Press, 1948).

Jaspers, Karl. *Man in the Modern Age* (New York, Anchor Books, 1957).

Jaspers, Karl. "Philosophy and Science," in *Way to Wisdom*, ed. Karl Jaspers (New Haven, Yale University Press, 1960), 147–167.

Jaspers, Karl. *Die Atombombe und die Zukunft des Menschen* (München, Piper, 1958), translated as *The Future of Mankind* (Chicago, The University of Chicago Press, 1961).
Jaspers, Karl. *The Question of German Guilt* (New York, Fordham University Press, 1965).
Jaspers, Karl. *Philosophy vol. 2* (Chicago, The University of Chicago Press, 1970).
Jaspers, Karl. *On Max Weber*, ed. John Dreijmanis (New York, Paragon House, 1989).
Kirkbright, Suzanne. *Karl Jaspers: Navigations in Truth: A Biography* (New Haven, Yale University Press, 2004), 208–214.
Masters, Dexter, and Katharine Way, eds., *One World or None: A Report to the Public on the Full Meaning of the Atom Bomb* (New York, New Press, 2007 [1946]).
Nehring, Holger. *Politics of Security: British and West German Protest Movements and the Early Cold War, 1945–1970* (Oxford, Oxford University Press, 2013).
Ruoppo, Anna Pia. *Marxismo ed esistenzialismo: due filosofie dell'Europa. Lukács e Jaspers si incontrano a Ginevra* (Milano, Mimesis, 2023).
Shklar, Judith. "Review of *The Future of Mankind*," *Political Science Quarterly* 76, no. 3 (1961), 437–439.
Spender, Stephen. "The Intellectuals and the Future of Europe," *The Gate/Das Tor* 1 (January/March 1947), 2–9.
Thornhill, Chris, and Ronny Miron. "Karl Jaspers," in *The Stanford Encyclopedia of Philosophy*, eds. Edward N. Zalta and Uri Nodelman (Summer 2024). https://plato.stanford.edu/archives/sum2024/entries/jaspers/
Weber, Max. "Politics as a Vocation," in *From Max Weber: Essays in Sociology*, ed. by H.H. Gerth and C. Wright Mills (New York, Oxford University Press, 1958).

4 Hiroshima Is Everywhere

Günther Anders' Reflection on the Atomic Threat

Micaela Latini

We are the masters of the Apocalypse. The infinite is us.

From the Human Being without a World to the World without a Human Being

Within the intellectual journey of the German thinker Günther Anders (born Stern, 1902–1992) lie the most significant milestones of the 20th century, but also (and above all) the crucial points of reflection on the necessity of becoming aware of the possible (or perhaps even certain) end of this very history. The dramatic succession of historical events during the 20th century (World War I, Hitler's rise to power with all the persecutions associated with it, the discovery of the horrors of Nazism, and finally the destruction of Hiroshima and Nagasaki) plays a crucial role in shaping Anders' philosophical thought.

If the rise of Nazism and the exile of millions of Jews (and non-conformists) represent a first dramatic point of reflection for the German thinker, who sketches the contours of a "human being without a world,"[1] the discovery of the Nazi concentration camps and the systematic and "administered" destruction of millions of people marks a "point of no return" and opens a previously unimaginable breach towards the Apocalypse. August 6, 1945, the date of the atomic bomb's detonation, represents for him the year zero, from which it no longer makes sense to focus on phenomenology, as all efforts must be directed towards the issue of the atomic threat.

Thus, his pen, like a sensitive seismograph, records the radical shift from the "human being without a world" to the "world without human being," which becomes the manifesto of his mature work. Anders' field of investigation, which had previously taken shape within a messianic horizon, now centers on the dimension of the "no more." In his short essay titled *My Judaism (Mein Judentum,* 1974), Anders, in opposition to Ernst Bloch's "not-yet," clearly distances himself from the dynamics of messianism:

I too, for many years, lived – in this, I was indeed very Jewish – in the expectation of the not-yet, of the establishment of the messianic kingdom. Until August 6, 1945 – the day of Hiroshima – when the idea struck me like lightning that perhaps, or even probably, we were heading towards a "no more." That was the end of my messianism. My task will now be this: to accept living without hope.[2]

The concept that defines the meaning of his mature work – or even the "fixed idea" (as Ernst Bloch, the philosopher of the "not-yet," polemically called it) – is the apocalyptic vision of a "world without life." The principle of despair replaces the principle of hope. Whereas previously Anders' focus was directed towards the scenario of estrangement and alienation of the modern human being in a "world-that-belongs-to-others" (a human being without a world), the awareness of the existence of new means of destruction at humanity's disposal opens up a very different ontological horizon: that of a potentially depopulated landscape and a humanity that remains alive only because "it has not yet been killed."[3]

A passage from the volume titled *The Man on the Bridge. Diary of Hiroshima and Nagasaki* fully captures the radical shift from "human being without a world" to "world without human being." We can read:

Hiroshima, August 6.
Chronology: August 6, 1945, is day zero. In fact, on that day it was proven that universal history may not continue, and that we are indeed capable of severing the thread of history; that day inaugurated a new historical era. A new era, though its essence lies in having a problematic existence.[4]

The dropping of the atomic bomb on Hiroshima marks a point of no return. It is the beginning of an era from which no God can save us, and in which the categories of possibility, hope, and future have become mere illusions, old keys that no longer open any doors.[5] August 6 represents "the day zero" of a new reckoning of time: the fracture between the "world of yesterday" and the present times lies in the fact that, for Anders, after that act, it is no longer possible to commit oneself to art, poetry, music, or literature. As he confesses in an interview: "When nuclear warheads are piling up, you cannot linger to explain the *Nicomachean Ethics*."[6]

The monstrous magnitude of the event first silences every form of muse. In Anders' words: "I understood immediately, already on August 7, one day after the Hiroshima attack and two days before the absolutely inexcusable one on Nagasaki, that August 6 marked the day from which humanity became irrevocably capable of self-destruction."[7] Of course, it took years before Anders was able to conceptualize such an extreme event, initially sketching, in a state of despair, while sitting under a walnut tree in the United States, a few lines about blindness towards the Apocalypse,[8] with the feeling of being unable to grasp the enormity of the situation.

The first pages of organized reflection on the nuclear issue are found in a crucial chapter of the first volume of his masterpiece *The Outdatedness of Human Beings* (1956), specifically in the section on the "roots of our blindness towards the Apocalypse" (which would be followed by a second volume in 1980, published under the title *On the Destruction of Life in the Age of the Third Industrial Revolution*).

The theoretical framework within which Anders reflects on atomic destruction is provided precisely by the question of technology. At the center of this study is the gap, the fundamental disparity between our "Promethean achievement" (the products made by us, the "children of Prometheus") and our ability to comprehend them. The issue at stake is a sort of "Promethean discrepancy" between producing (*herstellen*) and imagining (*vorstellen*).

We call "Promethean discrepancy" the ever-growing asynchrony between the human being and the world of his products, the distance that becomes greater with each passing day. Or in Anders' words:

> I call Promethean that difference according to the basic case of the gradient; that is, according to the gradient that exists between our "Promethean achievement," the products we manufacture as "children of Prometheus," and all other achievements: the fact that we are not equal to the "Prometheus within us."[9]

In the era of technocracy, explains Anders, the human being is no longer up to the level of his own products and even feels a sense of frustration, of shame, in the face of the unattainable perfection and power of the things that surround him, of his own creations.

A passage from *The Outdatedness of Human Beings* is revealing:

> To be a machine is precisely what he craves, it is precisely his task; the rattling robots that populate his cartoons are not, for him, figures without dignity or terrifying beings, but the embodiments of his dreams and his obligation to conform, disguised as puppets for the fun of him. […] If the human being suffers from a sense of inferiority in the face of his machines, it is primarily because, in his attempts to adapt to his devices and to make himself part of this or that machine, he must acknowledge that he constitutes a raw material of poor quality.[10]

Scientific thought and the products of technology offer the illusion of limitless development of human knowledge, but other faculties, those not strictly cognitive yet essential to processes of understanding, lag behind and risk turning out obsolete.[11]

Anders emphasizes the gap between doing and imagining, between acting and feeling, between knowledge and conscience, and finally, and most importantly, between the manufactured device and the human body (which is not built to the scale of the "body" of the device). All of these "gaps" share the same structure: one faculty is ahead of the other, thus one lags behind the other: just as ideological theory lags behind factual conditions, so does imagining lag behind doing.

In Anders' words:

> It is indeed true that we can make the hydrogen bomb, but we are not capable of imagining the consequences of what we have done. In the same way, our feelings lag behind our actions: with bombs, we can destroy hundreds of thousands of people, but we cannot mourn or regret them.[12]

What is at stake is a mechanism of "emotional illiteracy," that is, man's inability to measure the magnitude of his own actions and the loss of control over his own products once they are "thrown" into the world. In a long open letter to Klaus Eichmann, the son of the infamous Nazi party official, Anders introduces the theme of "inadequacy of feeling," due to the disproportion that separates our feelings from our actions, distancing us on one hand from the products we create or the actions we undertake, but on the other hand also distancing us from our emotions, leaving us "not even cold, but completely indifferent" precisely in the face of the "immense," the "too great," that which, though coming from the human, almost manages to place itself outside of us, in a separation from everything that constitutes an essential part of being human.

We read:

> What is too great leaves us cold, no, it doesn't even leave us cold, but completely indifferent. We become emotional illiterates who, faced with texts that are too vast, don't even realize they have them in front of their eyes. Six million to us remains just a number, while if we speak of ten murdered, perhaps something resonates within us, and a single murderer fills us with horror.[13]

The new form of emotional illiteracy that marks humanity is perfectly represented by the inexpressiveness of the means of destruction. As noted in an interview from 1979:

> Our perception is not equal to what we produce: how harmless the containers of Zyklon B gas seem – I saw them in Auschwitz – with which millions of men were destroyed! And how peaceful an atomic reactor looks with its domed roof!"[14]

And so, from the fateful date of August 6, 1945, Anders never ceases to question the nuclear issue, both through the publication of articles and books, and through active involvement as a co-promoter of the first anti-nuclear movements, with his journey to Hiroshima and Nagasaki, and not least, through the epistolary exchange that began in the late 1950s with Claude Eatherly, the young Texan meteorologist from the American reconnaissance plane who, on August 6, 1945, after observing sufficient visibility conditions, gave the go-ahead for the atomic bombing of Hiroshima. Let us begin with this epistolary exchange, published in 1961, to read it in continuity with the lines from *The Man on the Bridge. Diary of Hiroshima and Nagasaki*, written two years earlier.

The Beginning of the End Times

In 1959, by chance, Günther Anders, who had meanwhile moved to Vienna to follow his wife (the Austro-American writer Elisabeth Freundlich), came across an article in the American magazine *Newsweek* about the so-called pilot of Hiroshima. The article reported that the man had begun committing acts of petty crime and had therefore been confined to a military hospital. Anders, who (after turning down a university position at Halle offered by Bloch) had already established himself as a key figure in anti-militarist and anti-nuclear protests, immediately grasped the significance of this case and its exemplary nature.

It is worth recalling the circumstances that brought Eatherly into the spotlight. The young man, while flying over the islands in the reconnaissance plane "Straight Flush," determined that the skies over Hiroshima were clear, which meant good visibility of the island's primary target: the Japanese headquarters. But history – or perhaps in this case, the heavens – tragically intervened to return an enormous dose of responsibility to that day and to Eatherly's evaluation. He explained the disastrous events that followed his incorrect assessment in these terms: "The clouds over Hiroshima cleared and dispersed, the bombardier missed the target by about 3,000 feet, and destroyed the city of Hiroshima (...) a technical error that prevented the bomb from falling where it was supposed to."[15]

A human error, according to Eatherly's weak reconstruction. The result was that the device dropped on Hiroshima following his meteorological observation immediately caused the disappearance of 70,000 people; another seventy thousand died from radiation and burns in the following days. Three days later, hell was unleashed on Nagasaki with the dropping of the second atomic bomb, called "Fat Man": 39,000 people disintegrated instantly, and 25,000 more died in the weeks that followed, not to mention the effects of the atomic bombs on subsequent generations (on the children and fetuses of the few survivors).

That moment marked the beginning of the end times (*Endzeit*), an existence inherently threatened by the risk of a nuclear apocalypse, as we are reminded from time to time in alarming ways or "between the lines." While the pilots were celebrated and acclaimed back home at the end of the war as "smiling heroes," as bringers of peace, Eatherly did not recognize himself in this narrative. He withdrew into silence and devoted the rest of his life to trying to come to terms with his guilt and to making others aware of it.

In his conscience, the long shadow of the memory of his action in Hiroshima could not be easily set aside: the furies of his deed and the ghosts of the burning bodies on the bombed island began to haunt his sleep. Eatherly fell into a spiral of depression and attempted suicide several times. Under the weight of these conditions, his marriage to the Italo-American actress Concetta Margetti (whom he married in 1943) broke down, and he was barred from seeing his children.

In public life, too, he carried out a series of self-destructive acts of desperation: he broke into people's homes, committed petty theft, forged a check for a few dollars – in short, he tried to make himself guilty in the eyes of society. Through these antisocial acts, he sought to destroy the heroic image that Western society had made of him in order to continue justifying itself. However, Eatherly did not receive the punishment he sought; instead, he was declared mentally ill, diagnosed with a "recognizable personality disorder."

The figure of Eatherly, his story, immediately stood out to Anders as a symbol of humanity's condition in the age of weapons of mass destruction. For this reason, he decided to write him a letter:

> Dear Mr. Eatherly,
>
> You do not know the person writing these lines. However, you are known to us, to my friends and me. The way in which you will (or will not) come to terms with your misfortune is followed by all of us (whether we live in New York, Tokyo, or Vienna) with bated breath. And not out of curiosity or because we are interested in your case from a medical or psychological perspective. We are neither doctors nor psychologists. But because we are anxiously and earnestly trying to come to grips with the moral problems that confront all of us today. The technologization of existence: the fact that, indirectly and unknowingly, like the cogs of a machine, we can be involved in actions whose effects we cannot foresee, and which, if we could foresee them, we would not approve – this fact has transformed the moral situation for all of us. Technology makes it possible for us to become "innocently guilty" *[schuldos schuldig]* in a way that was still unknown to the technologically advanced world of our fathers. You understand your relationship to all this: because you are one of the first to have become entangled in this new kind of guilt, a guilt into which any one of us could fall – today or tomorrow.[16]

The letter continues with another interesting observation:

> You have the misfortune of having left behind two hundred thousand dead. And how would it be possible to experience grief that encompasses 200,000 human lives? Not only can you not do this, not only can we not do it; it is impossible for anyone. No matter how desperate the efforts may be, grief and remorse remain inadequate. The futility of your efforts is therefore not your fault, Eatherly; but a consequence of what I previously defined as the decisive novelty of our situation.
>
> [...] Because to be guilty as you are, and to be exalted for your guilt as a "smiling hero," must be an intolerable condition for an honest man; to put an end to it, one might even commit some transgressions. Because the enormity that weighed and continues to weigh on you was not understood, could not be understood, and could not be made understandable to the world to which you belong, you had to try to speak and act in an intelligible language [...] in terms recognizable to society itself, and so you tried to prove your guilt with acts that could be recognized as crimes. But even this did not succeed. You are always condemned to be seen as ill, rather than guilty.[17]

Anders' letter (which became the most famous of the correspondence) focuses on the issue of the Promethean gap experienced by human being in the technological society, a theme that runs like a thread through his philosophical reflection. The "discrepancy" between our capacity to produce and our ability to imagine has grown to such an extent that Eatherly is unable to make the event he materially caused accessible to the imagination, while the victims of that event do not fit within the scope of memory and must be removed.

Eatherly responded to the letter, and a correspondence developed between the two that lasted several months, during which Anders accompanied Eatherly on his journey, becoming almost a long-distance "philosophical advisor" *avant la lettre*. Eatherly thus became the subject of a philosophical experiment, a "philosophical guinea pig." Following Anders' suggestion, he began his journey with the reading of Augustine's *Confessions*; while interned in a mental health clinic, he turned to the works of Plato, Albert Schweitzer, and then immersed himself in the study of Anders himself, eventually attempting to write his own autobiography.

This dialogue has no consolatory purpose, nor does it intend to: "I have no intention of consoling you. Those who wish to console always say: 'It's not so bad'; in short, they try to diminish the event (whether grief or guilt) or to make it disappear with words. This is precisely what, for example, your doctors are trying to do."[18]

Rather, Anders tries to make Eatherly understand that what happened to him could, in some way, happen to any of us, that we are all potentially "pieces of an apparatus complicit in crime." The knowledge that humanity possesses in the age of technocracy is inadequate and entirely insufficient. For this reason, when faced with events of significant magnitude, even our feelings, our capacity to experience emotions, lag behind. From a philosophical perspective, for Anders, the "new phenomena of our time" necessitate the use of a different concept, that of a "supraliminal event" *[das Überschwellige]*: "I call 'supraliminal' – he clarifies – those events and actions that are too large to be comprehended by man: too large, therefore, to be perceived and remembered."[19]

The impossibility of imagining the functionality of today's technological products, their true scope, and their consequences relegates humanity to a marginal and obsolete position. Failing to "keep pace" with the work of its own products, the individual human being becomes a lever, a cog in the machine. For Anders, the issue is the "Promethean differential," the irreconcilable fracture between work and responsibility (a variation of the asynchrony between producing and imagining).

As Anders writes in the pages of the first volume of *The Outdatedness of Human Beings*:

> The faculties have grown apart and can no longer see each other; not seeing each other anymore, they no longer come into contact; and since they no longer come into contact, they no longer hurt each other. In short: the human being as such no longer exists, but there exists, on one hand, the one who acts or produces, and, on the other hand, the one who feels; man as producer and man as sentient being, and only these specialized fragments of men have a reality. What had filled us with horror ten years ago: the fact that the same man could be both a concentration camp worker and a good father, that the two fragments would not hinder each other because they no longer knew each other, this atrocious innocence of atrocity is no longer a singular case.[20]

It is a kind of "moral schizophrenia" in which a good father can be a criminal, and a criminal can be an employee. A striking example of this is the case of Adolf Eichmann, the banal man, the gray bureaucrat, the perfect official and perfect father, who was also a ruthless Nazi criminal.

As written in *The Outdatedness of Human Beings*, the worker in the extermination camp did not act, but paradoxically, he worked. Within this ethical framework, Eatherly represents for Anders the "antithesis" of Adolf Eichmann (1906–1962), the gray bureaucrat, free from responsibility, or the uncritical official who, by systematically producing corpses, simply followed orders.

In his own words:

But why should he repent, he who merely gave the go-ahead, who knew something about the discovery of nuclear fission only after his action, who simply obeyed orders and was merely used? Well, Mr. President, I could not accept this objection from anyone. I am Jewish, and I lost my friends in the Hitlerian gas chambers. With this excuse ("I simply obeyed orders"), all the officials involved in the extermination tried to justify themselves; and these words are all too eerily similar to Eichmann's, which we are reading in newspapers around the world these days: "In reality, I was nothing more than a small cog in the mechanism that carried out the directives and orders of the Reich. I am not a murderer nor a butcher" (*Life*, January 9, 1961). No, Eatherly is not Eichmann's twin, but his great and (for us) consoling antithesis. He is not the man who uses the mechanism as an excuse and justification for his lack of conscience, but the man who scrutinizes the mechanism as a fearful threat to conscience.[21]

After 15 years living under a false name in Argentina, Adolf Eichmann, without any apparent crises of conscience, declared himself not guilty before the Israeli court, claiming that as a "cog in the terror machine," he had merely obeyed orders. Eatherly, on the other hand, took a completely different path: he did not try to exonerate himself with the phrase "I was just a cog" in an inaccessible mechanism but felt the need to free himself from the role of a cog.

According to Anders, Eichmann's apparent impassivity during the trial in which he was a defendant was nothing more than a symptom of his "illness," his psychic-emotional blindness. Eatherly's case is different: in his desperate search to cope with the effects of his actions after the fact, he demonstrates that he kept his conscience alive within the technical and military machinery.

In 1964, Anders made another attempt at correspondence across the ocean, this time with Eichmann's son, Klaus Eichmann. But unlike Claude Eatherly, Eichmann did not respond to this message, nor to the second letter Anders sent in 1988. As a result, his letters, and thus the one-sided correspondence, turned into a single long letter of monologic reflection on "individual responsibility" in the technological world, published in 1964 under the title *We, the Sons of Eichmann (Wir Eichmannsöhne)*, which is interesting to read alongside Hannah Arendt's essay *Eichmann in Jerusalem*.

In Eatherly, Anders found a privileged interlocutor, a man willing to confront the horror of the demons he himself had summoned. The dialogue between the philosopher and the meteorologist managed to produce effects that no doctor or medicine had been able to achieve. This is a crucial point. In the face of such an obfuscation of consciences, of which Eatherly is the spokesperson, the method proposed by Anders consists of expanding our capacity to feel, broadening emotions through culture and imagination.

If the truth of our monstrous conditions is not fully perceptible, at least not to the naked eye, we must use imagination as a lens – a special lens capable of shrinking things, making them accessible to the limited human eye. It is imagination, indeed, that provides the condition for effectiveness in our lazy perception because it allows us to see a "vastness" that we ourselves are:

> our mere perception is insufficient to comprehend today's world, and it is too short-sighted for the enormous, or rather, monstrous dimensions of what we ourselves are capable of producing. [...] We must [...] use imagination as a corrective, as the truth of our monstrous conditions is certainly not perceptible, at least not to the naked eye [...] At the very least, we should be able to imagine [*vorstellen*] the vastness that we ourselves manage to produce [*herstellen*] and cause.[22]

The method proposed by Anders consists precisely of distortion, of exaggeration, which should not be understood as a falsification of the truth. Rather, it is a method of observation, of grasping at least the truth content of the lie and communicating it by expanding our capacity to feel (without necessarily inducing a specific emotion). According to Anders, this is what culture must do, at least after Auschwitz and Hiroshima: it must assault the limits (to use Kafka's words) by broadening emotions through literary or philosophical works, films, or music.

Only in this way can the human being attempt to awaken and re-educate his imagination, reconnect with his emotions, and finally learn to despair. For these reasons, Anders can advocate and defend the importance of the principle of despair: "And thank God they are now despairing, at last, they are despairing.[23]

Thus came the decision to publish the correspondence, which was first released in 1961 in German under the title *Off-Limits für das Gewissen*, with an introduction by Robert Jungk, who in 1956 had authored the now-canonical *The Sorcerer's Apprentices (Heller als tausend Sonnen,* 1956). Upon its release, the Anders-Eatherly correspondence garnered worldwide attention. The figure of Eatherly and his story captured public attention. In English-speaking countries, the book was released in 1962 under the title *Burning Conscience,* with a foreword by Bertrand Russell. It is worth noting that in the same year, 1956, the German philosopher Karl Jaspers – who had already engaged in a specific correspondence on the subject of the "nuclear bomb" with Hannah Arendt – delivered a controversial radio address titled *The Atomic Bomb and the Future of Human Being (Die Atombombe und die Zukunft des Menschen,* 1956), in which he adopted pacifist positions while also showing some alignment with US policies. The nuclear debate was indeed a topical issue, not only in the United States. However, unlike Jaspers, Anders avoids dwelling on political nuances he deems irrelevant, which only serve to obscure the true threat he identifies: technology left unchecked. As the Austrian journalist Robert Jungk aptly noted in the German introduction to Anders' work (1961), the destructive violence of atomic weapons transcends all previous wartime experiences. The psychological repercussions on those who wield atomic weapons cannot be processed either consciously or unconsciously. Eatherly, burdened by guilt for his role in the events he helped unleash, becomes a silent scream, which Anders chooses to amplify. That scream concerns us all, as this story is our story, revealing our collective encounter with something too overwhelming: the risk of a nuclear apocalypse.

Apocalyptic Blindness

In June 1958, Anders participated in an international anti-nuclear armament conference in Japan, where he drafted an interesting diary later published under the evocative title *Der Mann auf der Brücke (The Man on the Bridge)*. His message in the introduction to this work is clear. Those seeking to rationally confront the atomic situation must undergo a "Copernican revolution," internalizing a previously unfamiliar axiom: "since our world stands at a 'to be or not to be' crossroads due to atomic weapons, the world situation is defined by the atomic fact."[24] Through technological progress, modern humanity has irrevocably initiated – like Goethe's sorcerer's apprentice – an endgame scenario, where we scramble to checkmate ourselves. Goethe's *Zauberlehrling* (sorcerer's apprentice) appears prominently in the second volume of *The Outdatedness of Human Beings*, symbolizing modern humanity. However, unlike the apprentice who hoped for a happy ending, today, as Anders forewarns, we cannot count on one.

In light of the nuclear threat, there is no turning back: nuclear weapons might be eliminated in an ideal world, but the capacity to create them remains. In the *Diary of Hiroshima and Nagasaki,* Anders notes,

> What has happened is irrevocable precisely because it can always be repeated. We may abolish atomic weapons, but we cannot eliminate the knowledge to produce them. Our technical situation is defined not only by what we control but also by what we are incapable of not controlling.[25]

Thus, Anders explains, we cannot delude ourselves into "living without the bomb." Yet, nor can we accept the opposite view – "living with the bomb" – a grim and deeply ambiguous formula coined by physicist Carl Friedrich von Weizsäcker, who perhaps unintentionally, lent support to proponents of rearmament. Anders argues that the formula is not entirely dismissible, as it conveys a profound truth: "we must not deceive ourselves, not even for a moment, into believing we can live 'without the bomb,' even in a future where bombs no longer exist."

In the *Diary of Hiroshima and Nagasaki*, Anders focuses on the metamorphosis of war, which has taken on a "phantom-like and indirect" character, becoming a paradoxical and undefined phenomenon.[26] The era of atomic threat presents an entirely different scenario from past world wars. As Anders writes, "Those were peaceful times when there were still wars!"[27] And further: "Back then, soldiers massacred each other and fought wars as people capable of hatred. Those who hated each other could one day stop hating, or even learn to love."[28]

Not only has the face of war changed, but what we now call war – a potentially atomic conflict – cannot be compared to conventional war. "War can only be discussed when military action leaves a continuous life stream; when it assumes an enemy capable of defending itself; when it aims at a state of affairs post-war that is not nullified by the act itself."[29]

Throughout his work, Anders argues that the image of the soldier recognizing his enemy by uniform and aiming at him is obsolete. The enemy has become faceless, and hatred has vanished along with it. Bombs have no eyes, enemies have no faces, and battlefields no longer exist. Artillerymen, unable to identify their targets, are unaware of their actions. Even the location of the crime has disappeared: in the "schizopathic" nature of modern warfare, a missile launched in the Pacific Ocean may have its impact in Siberia.

In Anders' words:

> *Bombs and missiles have no eyes to distinguish uniforms from civilian clothes; today's égalité lies in the fact that every civilian has the same right as a soldier to be killed [...] to the executors, everything is identical, everything has the same value, everything is indifferent; their actions are absolute indifference, nihilism in action.*[30]

We must focus our attention on the framework of the "new" post-atomic situation, in which all forms of distance have disappeared: the "fall-out" has once and for all dismantled the coordinates of near/far. Anders' study immediately highlights the breakdown of space and time concepts. The point is that the idea of topography has been eliminated in the nuclear environment, as the atomic bomb is omnipresent. "We are not only contemporaries, but also *cospatial*, inhabitants of the same space."[31] Similarly, the fact that humanity has produced this tool of destruction renders it omnipotent, yet, in a dramatic reversal, also powerless.

Anders explains how, in the atomic state, there are no borders and no "civil wars." He states,

> With the disappearance of borders brought about by the atomic situation, this distinction [between war and civil war] has also vanished ... The second thing we learned from your experiences is that what are called tests today are not tests at all, nor experiments.[32]

They are actual atomic explosions. At the same time, the attempt to create so-called "clean" bombs in place of "atomic weapons" in general appears absurd. Even the concept of "weapons" is entirely inadequate and misleading in this context: a weapon is still a means in the traditional sense, used to achieve objectives; the bomb, however, serves no purpose other than "total destruction." If Anders repeatedly and persistently urges us not to call these objects "weapons," it is because "the magnitude of their effects inherently surpasses any possible purpose, and their use cannot be justified by any conceivable objective."[33]

In truth, even the notion of peace is outdated, reduced to a "continuation or preparation of war by other means." And this is to be expected, given that war, since Hiroshima, is no longer what it once was. Anders provocatively asks: "Was there still resistance? Could there be resistance?"[34]

His task, therefore, is to bring realism back to the scene, to make the invisible visible in a paradoxical way, by reversing the perspective. It is about encouraging the eyes to close in order to see better, to escape the state of "apocalyptic blindness." In his words:

> Whoever today limits themselves to perceiving what is immediately visible misses out on reality. For today's reality is pregnant with the fantastic. Whoever lacks imagination or prevents themselves, out of fear, from fully grasping the fantastic remains a dreamer. The so-called "perceptual world" is an ivory tower, and empiricism an escapist haven. – To truly see today, one must close their eyes; and a realist today is only someone with enough imagination to envision the fantastic tomorrow.[35]

This is also what Anders asserts in another entry from his diary, dated 1979, recounting his impressions from his non-return journey in Breslau. The volume, published as *Descent into Hades (Besuch im Hades)*, addresses the theme of the Shoah:

> The imagination required today no longer means what we once understood by that term, no longer involves exaggerating the real or envisioning unreal beings or fantastic creatures ... On the contrary, today imagination means confronting today's truly fantastic reality and interpreting it appropriately. In sum: since its object, the fantastic reality, is itself fantastic, imagination must function as a method of empiricism, as a perception organ for the genuinely immense.[36]

"Imagination as a method of empiricism": this means that for Anders, art, as a tool for action and knowledge, still represents a privileged perspective for understanding and even transforming reality. Indeed, Anders sees imagination as the only moral organ relevant to truth and ethics.[37]

The point is that the real does not correspond to the visible but is created by imagination. Therefore, his advice to the traveler can be summed up in this formula:

> Stay here and walk on the roads and bridges. And think where you walk, and on what, and who you are. And consider that nothing you see is real; that real is only the fact that you no longer see reality, that you can no longer see it. Close your eyes and trust your imagination. For only the lazy rely on their eyes today.[38]

In line with the observations made by Benjamin in *Erfahrung und Armut (Experience and Poverty)* regarding World War I, Anders also emphasizes the poverty of experience that the wartime event entails, which is closely linked to the development of technology. The point is that we are unable to imagine beyond what we perceive; we cannot think of what we are capable of producing. Thus, Anders urges us to exercise our imagination and creativity because we are reverse utopians. While utopians fail to produce what they imagine, we struggle to imagine what we ourselves have produced.

But there is more. We are incapable of representing catastrophe because we cannot conceive of that total abstraction that is non-being, the absence of the human world. It is something too immense to fit into the conceptual categories we have available:

> The victims (strange as it may seem) did not "live" the catastrophe itself, but only life before it and life (or death) after it. Not the flash in between. That was something too immense, arriving and disappearing too instantaneously for them to grasp and conceive of it as such. This inconceivability persists even now; for even now they cannot express the event in words or give it a name, tending to circumscribe it with an "it" ("then it happened"). Or perhaps it is a euphemism.[39]

In perfect analogy with the horror of Auschwitz, what happened in Hiroshima and Nagasaki also escapes a comprehensive definition. Based on this "*Augenblick*" that eludes every glance, Anders deduces the invisibility of the coming end: in the era of intercontinental missiles, the enemy will no longer be seen, the weapon will no longer be seen, the strike will no longer be seen. It will only happen that "ruin will descend upon the moment and kill everything, and even the effect of the blow will no longer be seen."[40]

To Be and Not to Be

The question of non-being, of nothingness, constitutes a critical point in Anders' text. But not in the abstract sense of Sartre, rather in the form of a concrete place. Hiroshima represents the peak of nihilism reached by humanity, a milestone that was unimaginable in yesterday's world and that suddenly tore through the veil of everyday life. The stakes are "the thing itself, the Nothing in itself, the Nothing for adults: annihilation, massive, total physical destruction, leaving nothing that has not been destroyed."[41]

The problem lies in the fact that this Nothing is nowhere, is in no moment, yet at the same time is present everywhere and at every instant, thus able to upend every apparent compactness of the real.[42]

It is no coincidence that many pages of Anders' *Diaries* are dedicated to the issue of monuments, or rather to the memory of the tragedy that struck the cities of Hiroshima and Nagasaki. More than monuments, it would be appropriate in these cases to speak of "nonuments," that is, memorials that, built to preserve memory, paradoxically function as machines of oblivion.[43]

How is it possible to convey the horror of the nuclear Holocaust? Anders' reflections are dedicated to the reconstruction of the city, a reconstruction that is also "the destruction of destruction," and thus represents the culmination of destruction itself:

> As I understand the sense of estrangement that O. speaks of! For his impression of "not being where he is" is an experience so familiar to all of us inhabitants of the rebuilt cities that it has almost lost its terrifying character [...]. Yes, when I saw the city for the first time again, when it was still a heap of ruins, it was still my city; but now![44]

The fact is that the new (things and houses) returns a halo of atemporality, as if it had always been this way. The past is thus masked, camouflaged, and history, which has produced the very falsehood, reveals itself as the author of the false: "History as the history of its own falsification."[45] Within this discourse on the paradox of memory, Anders examines the Hiroshima memorial: a concrete arch that does not resemble a "tower" at all, a *Peace Tower,* but rather a bridge, which, however, does not lead anywhere. Moreover, it is precisely on the bridge in Hiroshima that one finds a man, or rather a composite of being and non-being, almost an automaton.[46]

> Initially, we heard only the fragile sound of the strings plucked by an instrument. But then we "saw" it. "It": a person or a thing? This uncertainty already tells you everything. What was playing down there was, at first glance, unrecognizable. A curtain hung from the head, preventing us from seeing the face. The reason for this was all too clear for a visitor to Hiroshima.
>
> The gaze slid along the body (for what contributes most to making a man a man, after the face, are, of course, the hands). But no hands were visible. What was playing was a thing of steel. Even the reason for this was all too clear for a visitor to Hiroshima.[47]

What is on the bridge is, therefore, an automaton. And what scares most about it is not the lack of face and hands, but that between the two devastated and dehumanized regions that were once called "face" and "hand," there stretches a real, living, and pulsating body. This is inconceivable.[48]

At the center of these observations is the bare and pulsating life of the man from Hiroshima, his body, his being, stretching like a cord between the non-being of hands and face. This man, balancing between being and non-being, represents what remains of Hiroshima. But at the same time, Hiroshima, an emblem of nothingness, is everywhere.

Hiroshima Is Everywhere

In the final section of his research, Anders' reflection on nuclear issues radicalizes, intersecting with questions of violence. The positions expressed by the author, who had always been seen as a champion of nonviolent struggle, scandalize many. Man has proven capable of exterminating, with the mere press of a button, all of humanity, and this form of self-extinction can be realized in complete unawareness, exceeding the limits of human understanding. From this premise, from the increasingly tangible risk of catastrophe, derives, according to Anders, the state of necessity. With the tragedy of the nuclear explosion at Chernobyl – Anders explains – the abstract threat has become concrete, and for this reason, those who support nuclear power can be considered a source of terrorism. Radioactive contamination is invisibly dangerous and, in a note of eternity, will remain active, triggering a sort of epidemic that can reach all points of the earth.[49] In this sense, the dramatic sentence "Chernobyl is everywhere" must be considered even stronger than "Hiroshima is everywhere."[50]

The state of necessity is due to the fact that, after the tragedy of Chernobyl, we live under the sword of Damocles: with the constant threat of a nuclear explosion. Anders has no doubts: the pre-revolutionary stage of our sentimental and symbolic protest against the preparation for total annihilation now belongs to the past. Those who today promote and support the construction of weapons of mass destruction place us in a state of global emergency. We must then begin to attack and neutralize these monsters of nuclear destruction. In short, the demons that crowd the world – and endanger all humanity – are precisely those who advocate for power plants. They are the ones who have reduced the state of their time frame to a state of exceptional necessity:

> We find ourselves in a situation that, legally speaking, is a state of necessity [*Notstand*]. In all codes, even the Canon Law, in a situation of state of necessity, violence [*Gewalt*] is not only permitted but recommended.[51]

There is no doubt: this is an extreme response, which perhaps must be followed in all its reasoning. In a sort of crossed step between production and destruction, human beings have created weapons of mass destruction and have thus condemned themselves to a "state of necessity": millions of men are threatened with death, not by other men who want to kill other men, but by those who take the risk and can think only technically and factually. Unraveling a long thread of reasoning, Anders arrives at the necessity of counter-violence, of an act capable of containing, even aggressively, the actions of those who terrorize all of humanity. In reality, the words uttered by Anders sometimes reach extreme heights:

> There are no alternatives but the threat if we do not want to hope for the survival of ourselves and future generations. We must communicate to these people that they will be considered from now on, one by one, as prey, as game. They become people that anyone could kill.[52]

A few pages later, in the same text, Anders returns to the same issue, emphasizing how resorting to violence, while not fully legitimized from a legal point of view, is, under certain assumptions, a cornerstone of morality, and for this reason, above legality. On this basis, he advocates the necessity, even for those who have always embraced pacifism, to revisit their positions:

> One cannot and must not become or be or remain at all costs an advocate of non-violence because anyone who is threatened and attacked – and this is provided not only by international law but also by Canon law – is authorized and even obliged to legitimate defense against threats of violence and more than ever against acts of violence (…). Therefore, we have the right to exercise counter-violence, even if this too cannot rely on any administrative or legal power, in short, on no state. But the state of necessity (*Notstand*) legitimizes self-defense (*Notwehr*), morality wins over legality.[53]

In the opening of this statement, it almost seems that Anders is addressing himself, correcting and updating his previous statements. This position, that of violence as a possible salvation, has shocked many of his admirers, convinced of Anders' pacifist faith. The convictions expressed in *Gewalt. Ja oder nein* must have been received by Andersians as a boulder in the stomach, given that the philosopher had always stood out as a spokesperson for pacifist currents. But why this distancing from the reasons of pacifism tout court? Anders is keen to emphasize that this is a "natural" transition, a rectification for those who instead define it as a "change of course."

In the new conditions in which humanity finds itself, it is necessary to "update" the old slogans, re-evaluating them in light of the changes that have occurred, and of the new "state of things." In his view, the error of many representatives of pacifist movements lies in having forgotten that peace is not a means, but an end:

But for me, peace is not a means, but a goal. [...]. I can no longer stand by and watch those of us who, together with our descendants, are mortally endangered by the violent, who shy away from using violence against the threat of violence.[54]

Only by keeping in mind the *Wozu* is it possible to restore the citizenship visa to "realistic pacifism," that is, to that form of conduct that contemplates violence to stop further violence. In the fabric of our time – Anders explains – the traditional concept of pacifism should be considered outdated, superfluous, since there is no longer any alternative to being pacifists.[55] Likewise, violence is reconsidered as necessary for the pursuit of pacifist ends, a violence aimed at overcoming violence (*Gewalt also zwecks Aufhebung von Gewalt*).[56] Hence, Anders comes to favor the violence of self-defense, in the event that our very existence is threatened:

The right to self-defence for those who are threatened with death and can be attacked at any moment is, of course, natural! (...) And since the threat is global and the possible extermination is global, our legitimate defence must also become total and global.[57]

Anders' position is as clear as it is extreme: today there is no method of salvation except threatening those who threaten us. For this reason, he declares open war on those who have forced humanity to break the taboo of not killing:

In short, we must consider enemies and treat as such all those who (as happened with Hitler's war, but also with that of Kennedy and Johnson in Vietnam) *force us to do what is truly taboo for us: kill.* To the commandment "*You shall not kill!*" (Exodus 20:13), which is now over 3,000 years old, we should add an integration: "*You are allowed to kill, or perhaps even: you must kill those who are ready to kill humanity, and who demand that other men, thus us, approve their threats and participate in their actions.*"[58]

A similar thesis had already been anticipated by Anders a few years earlier in a passage from his book *Ketzereien* (1982). On that occasion, he condemned the commandment "You shall not kill" as a morally dogmatic principle that is not too defensible and that should be broken.

We read in a passage from the same volume, where he shifts his focus to what he considers the new Holocaust, namely the nuclear disaster:

[...] if today there were a politician – I say, if – who wrote a book suitable for Hitler, and who thus proclaimed the existence of a good superior to universal peace, and who expressed favourably regarding his willingness to risk nuclear war, no: who even just implied this willingness [...] then I would be for [...] killing him.[59]

In front of an embarrassed and surprised interlocutor at these extreme statements, Anders clarifies the reasons for violence: aggressiveness is legitimate in certain circumstances and if directed against those who attack the humanity of man. Anders' first thought goes to the tragedy of the Holocaust, and – as we know from the text on violence – to his Austrian writer friend Jean Améry (Hans Mayer), who before dying confessed his regret for never having threatened to kill those who hunted him for life.

In response to the question that these notes could be interpreted as favoring terrorism, he states that terrorism was represented by the bombing of Hiroshima and the preparation for nuclear war and argues that, in comparison, those who might occupy a factory or, to put it more softly, conduct a terrorist attack should be considered innocent. He also emphasizes that Mr Truman's good conscience did not make the crime he committed any less serious.[60]

For the last Anders (who died in Vienna in December 1992), violence is therefore permitted if it completely coincides with that counter-violence that is legitimate defense: "*the exercise of counter-violence to which we are forced is legitimate solely because it aims to create the situation of non-violence.*"[61] One could say in almost paradoxical terms that Anders, for the love of pacifism, must boycott pacifism itself.[62] This position is undoubtedly extreme, but it fully responds to that art of exaggeration that the thinker has always practiced, which earned him the nickname "Cassandra of philosophy." But beware: this is a hyperbolic method to make one's voice heard or to open the eyes of those who continue, incessantly, to support and produce a dangerous form of blindness in the face of the Apocalypse.

Notes

1. Under this dictum are collected texts such as *Mensch ohne Welt. Schriften zur Kunst und Literatur* (Munich, Beck Verlag, 1984), as well as "Pathologie de la liberté. Essai sur la non-identification" (in *Recherches Philosophiques* 1936–1937, vol. 6, 22–54. English translation by K. Wolfe, "The Pathology of Freedom: An Essay on Non-Identification," *Deleuze Studies* 3, no. 2 (2009), 278–310. K. Wolfe, *Kafka: Pro und Contra. Die Prozeß-Unterlagen* (Munich, Beck Verlag, 1951).
2. G. Anders, *Mein Judentum,* hrsg. von H. J. Schultz (Stuttgart, Kreuz Verlag, 1978), 59 (our translation).
3. G. Anders, *Hiroshima ist überall: Tagebuch aus Hiroshima und Nagasaki. Der Briefwechsel mit dem Hiroshima-Piloten Claude Eatherly. Rede über die drei Weltkriege* (Munich, Beck Verlag, 1995), 66 (our translation).
4. G. Anders, *Der Mann auf der Brücke*, 86 (our translation).
5. The metaphor is taken from G. Anders, *Der Blick vom Turm. Fabeln. Mit 12 Abbildungen von A. Paul Weber* (Munich, Beck Verlag, 1968), 83–84.
6. G. Anders, "Wenn ich verzweifelt bin, was geht's mich an?" In *Die Zerstörung einer Zukunft. Gespräche mit emigrierten Sozialwissenschaftlern,* hrsg. von M. Greffrath (Hamburg, Rowohlt, 1979), 46 (our translation).

7 G. Anders, "Wenn ich verzweifelt bin, was geht's mich an?" 42 (our translation).
8 G. Anders, "Wenn ich verzweifelt bin, was geht's mich an?" 45.
9 G. Anders, *Die Antiquiertheit des Menschen. Band I: Über die Seele im Zeitalter der zweiten industriellen Revolution* (Munich, Beck Verlag, 1961), 270 (our translation).
10 G. Anders, *Die Antiquiertheit des Menschen. Band I*, 43, 49–50 (our translation).
11 Anders touches on this issue throughout much of the book.
12 G. Anders, *Die Antiquiertheit des Menschen. Band I*, 17 (our translation).
13 G. Anders, *Wir Eichmannsöhne. Offener Brief an Klaus Eichmann* (Munich, Beck Verlag, 1988), 28–29 (our translation).
14 G. Anders, "Wenn ich verzweifelt bin, was geht's mich an?" 43 (our translation).
15 G. Anders, *Hiroshima ist überall*, 294 (our translation). Cfr. also *Burning Conscience. The Case of the Hiroshima Pilot, Claude Eatherly, Told in His Letters to Günther Anders* (New York, Monthly Review Press, 1962), 81.
16 G. Anders, *Off Limits für das Gewissen, der Briefwechsel zwischen dem Hiroshima-Piloten Claude Eatherly und Günther Anders*, hrsg. von R. Jungk (Hamburg, Rowohlt, 1961), 17 (our translation). Cfr. also *Burning Conscience*, 1.
17 G. Anders, *Off Limits für das Gewissen*, 17 (our translation). Cfr. also *Burning Conscience*, 3–5.
18 *Burning Conscience* 2.
19 G. Anders, "Wenn ich verzweifelt bin, was geht's mich an?" 45 (our translation).
20 G. Anders, *Die Antiquiertheit des Menschen. Band I*, 272 (our translation).
21 *Burning Conscience*, 108–109.
22 G. Anders, *Besuch im Hades. Auschwitz und Breslau 1966. Nach «Holocaust» 1979* (Munich, Beck Verlag, 1979), 40 (our translation).
23 G. Anders, *Besuch im Hades*, 202 (our translation).
24 G. Anders, *Hiroshima ist überall*, 6 (our translation).
25 G. Anders, *Hiroshima ist überall*, 23 (our translation).
26 G. Anders, *Hiroshima ist überall*, 262 (our translation).
27 G. Anders, *Hiroshima ist überall*, 76 (our translation).
28 G. Anders, "Die Antiquiertheit des Hassens," in *Haß. Die Macht eines unerwünschten Gefühls*, hrsg. von R. Kahle u. a. (Hamburg, Rowohlt, 1985), 28.
29 G. Anders, *Hiroshima ist überall*, 76 (our translation).
30 G. Anders, *Die Antiquiertheit des Hassens*, 30 (our translation).
31 G. Anders, *Hiroshima ist überall*, 37 (our translation).
32 G. Anders, *Hiroshima ist überall*, 38 (our translation).
33 G. Anders, *Hiroshima ist überall*, 37 (our translation).
34 G. Anders, *Hiroshima ist überall*, 38 (our translation).
35 G. Anders, *Hiroshima ist überall* 48 (our translation).
36 G. Anders, *Besuch in Hades*, 39 (our translation).
37 G. Anders, *Besuch in Hades*, 39.
38 G. Anders, *Hiroshima ist überall*, 66 (our translation).
39 G. Anders, *Hiroshima ist überall*, 85 (our translation).
40 G. Anders, *Hiroshima ist überall*, 84 (our translation).
41 G. Anders, *Hiroshima ist überall*, 80 (our translation).
42 G. Anders, *Hiroshima ist überall*, passim.
43 Cfr. A. Pinotti, *Nonumenti. Un paradosso della memoria* (Milan, Johan & Levi, 2023).
44 G. Anders, *Hiroshima ist überall*, 62 (our translation).
45 G. Anders, *Hiroshima ist überal*, 62 (our translation).
46 It is worth remembering that Luigi Nono composed the *Canti di vita e d'amore: Sul ponte di Hiroshima* in 1962 for soprano, tenor, and orchestra, with texts by Günther Anders taken from "Der Mann auf der Brücke," in *Hiroshima ist überall: Tagebuch aus Hiroshima und Nagasaki*.

47 G. Anders, *Hiroshima ist überall*, 69 (our translation).
48 G. Anders, *Hiroshima ist überall*, 69.
49 G. Anders, "Zehn Thesen zu Tschernobyl," *Taz* (3 June, 1986), 8.
50 It is no coincidence that the texts on Hiroshima were later published in the volume with the title *Hiroshima ist überall*.
51 G. Anders, *Gewalt – Ja oder Nein? Eine notwendige Diskussion*, hrsg. von M. Bissinger (Munich, Knaur, 1987), 23 (our translation).
52 G. Anders, *Gewalt – Ja oder Nein?*, 92.
53 G. Anders, *Gewalt – Ja oder Nein?* 93 (our translation).
54 G. Anders, *Gewalt – Ja oder Nein?*, 108 (our translation).
55 G. Anders, *Gewalt – Ja oder Nein?* 90.
56 G. Anders, *Gewalt – Ja oder Nein?* 101.
57 G. Anders, *Gewalt – Ja oder Nein?*, 91 (our translation).
58 G. Anders, *Gewalt – Ja oder Nein?*, 145 (our translation).
59 G. Anders, *Ketzereien [Heresies]* (Munich, Beck Verlag, 1982), 337 (our translation).
60 See K. P. Liessmann, *Günther Anders zur Einführung* (Hamburg, Junius Verlag, 1988), 158 (our translation).
61 G. Anders, *Gewalt – Ja oder Nein*, 103 (our translation).
62 G. Anders, *Gewalt – Ja oder Nein*, 90.

Bibliography

Anders, Günther. "Pathologie de la liberté. Essai sur la non-identification," *Recherches Philosophiques* 6 (1936–1937), 22–54. English translation by K. Wolfe. "The Pathology of Freedom: An Essay on Non-Identification," *Deleuze Studies* 3, no. 2 (2009), 278–310.

Anders, Günther. *Kafka: Pro und Contra. Die Prozeß-Unterlagen* (Munich, Beck Verlag, 1951).

Anders, Günther. *Die Antiquiertheit des Menschen. Band I: Über die Seele im Zeitalter der zweiten industriellen Revolution* (Munich, Beck Verlag, 1961).

Anders, Günther. *Off Limits für das Gewissen, der Briefwechsel zwischen dem Hiroshima-Piloten Claude Eatherly und Günther Anders*, hrsg. von R. Jungk (Hamburg, Rowohlt, 1961).

Anders, Günther. *Burning Conscience: The Case of the Hiroshima Pilot, Claude Eatherly, Told in His Letters to Günther Anders* (New York, Monthly Review Press, 1962).

Anders, Günther. *Der Blick vom Turm. Fabeln*. Mit 12 Abbildungen von A. Paul Weber (Munich, Beck Verlag, 1968).

Anders, Günther. *Mein Judentum*, hrsg. von H. J. Schultz (Stuttgart, Kreuz Verlag, 1978).

Anders, Günther. "Wenn ich verzweifelt bin, was geht's mich an?" In *Die Zerstörung einer Zukunft. Gespräche mit emigrierten Sozialwissenschaftlern*, hrsg. von M. Greffrath (Hamburg, Rowohlt, 1979).

Anders, Günther. *Besuch im Hades. Auschwitz und Breslau 1966. Nach «Holocaust» 1979* (Munich, Beck Verlag, 1979).

Anders, Günther. *Ketzereien* (Munich, Beck Verlag, 1982).

Anders, Günther. *Mensch ohne Welt. Schriften zur Kunst und Literatur* (Munich, Beck Verlag, 1984).

Anders, Günther. "Die Antiquiertheit des Hassens," in *Haß. Die Macht eines unerwünschten Gefühls*, hrsg. von R. Kahle u. a. (Hamburg, Rowohlt, 1985).

Anders, Günther. "Zehn Thesen zu Tschernobyl," *Taz*, June 3, 1986.

Anders, Günther. *Gewalt – Ja oder Nein? Eine notwendige Diskussion*, hrsg. von M. Bissinger (Munich, Knaur, 1987).

Anders, Günther. *Wir Eichmannsöhne. Offener Brief an Klaus Eichmann* (Munich, Beck Verlag, 1988).

Anders, Günther. *Hiroshima ist überall: Tagebuch aus Hiroshima und Nagasaki. Der Briefwechsel mit dem Hiroshima-Piloten Claude Eatherly. Rede über die drei Weltkriege* (Munich, Beck Verlag, 1995).

Liessmann, Konrad Paul. *Günther Anders zur Einführung* (Hamburg, Junius Verlag, 1988).

Pinotti, Andrea. *Nonumenti. Un paradosso della memoria* (Milan, Johan & Levi, 2023).

Index

absolute justice 39
Acheson, Dean 9
Acheson-Lilienthal Report 9
age of deterrence 30–34
aggression, Jaspers' view of 53
Améry, Jean 81
Anders, Günther 3–4, 63–67;
apocalyptic blindness 73–76;
atomic bombs 63–67; borders
74–75; correspondence with
Eatherly 68–73; counter-
violence 79; emotional illiteracy
66; emotions 72; Hiroshima
memorial 77; imagination
75–76; inadequacy of feeling
66; non-being 76–78; Nothing
76–78; nuclear weapons 73–76,
78–81; pacifism 81; peace 75;
Promethean discrepancy 65;
psychic-emotional blindness 71;
supraliminal events 70; violence
79–81
Angell, Norman 20
Anscombe, Elizabeth 3
anti-nuclear movements 67, 73
apocalypse 58, 64–65
apocalyptic blindness, Anders' view
of 73–76
Arendt, Hannah 48–49, 52, 71, 72
Aron, Raymond 3–4, 26–27, 38–40,
45; age of deterrence 30–34;
fanaticism 39; nuclear strategy
34–38; nuclear weapons 29–30;
peace 28–30; war 28–30
Association of Scientific Workers 8
Atlantic alliance 37–38

atomic bombs: Anders' view of
63–67; Jaspers' view of 44–45,
57–58; Russell's views on 7–8,
12–14; totalitarianism 51–54;
see also Hiroshima; Nagasaki
atomic consciousness 2
Atomic Development Authority 9
atomic equality 30
Atomic Scientists' Association 8
atomic unconsciousness 2
Attlee, Clement 8
Augustine 69

Baruch, Bernard 9
belonging, national belonging 10
Benjamin, Walter 76
Bevin, Ernest 8
Bikini test 11–12
Blücher, Heinrich 49
Bobbio, Norberto 3
Bohr, Niels 7, 9
borders, Anders' view of 74–75
boundary situations, Jaspers' view
of 46, 57–58
Bulganin, Nikolai Aleksandrovich 14
*Burning Conscience (Off-Limits für
das Gewissen, Anders)* 72, 82
Byrnes, James 9

Campaign for Nuclear
Disarmament 16
capitalism 52
change: Jaspers' view of 54–55;
Russell's views on 8–9
Chernobyl 4, 78
Chinese civil war 31

cleavages 10
Cold War 28–29, 56–57
collective self-glorification 18
collective suicide 58
colonialism 53
Common Sense and Nuclear Warfare (Russell) 16–17
communication 54
conscientious objectors 6
contractualism 48, 54–55
Copernican revolution 73
correspondence between Anders and Eatherly 68–73
Coty, René-Jules-Patrice-Gustave 14
counter-violence, necessity of 79
Cousins, Norman 8
Cripps, Stafford 8
cultural cleavage 10

decolonization 53
de Gaulle, Charles 37–38
democracy, Jaspers' view of 55–56
democratization 29
Der Blick vom Turm (Anders) 81
Descent into Hades (*Besuch im Hades,* Anders) 75
despair 64
despotism 52
destruction 77
détente 14, 17
deterrence 30–34
Dickinson, Goldsworthy Lowes 20
Die atomare Drohung (Anders) 4
Die Atombombe und die Zukunft des Menschen (Jaspers) 44
Dulles, Foster 18, 34

Eatherly, Claude 67–71
Eden, Anthony 14
education, Russell's views on 18–19
Eichmann, Adolf 66, 70–71
Einstein, Albert 14–15
Eisenhower administration 34, 37
Eisenhower, Dwight 14–15
emotional illiteracy, Anders' view of 66
emotions 72
Enlai, Zhou 14

equality, Jaspers' view of 45
escalation 33, 36
ethical idea, Jaspers' view of 50–51
Europe, nuclear strategy 35–36
European anxiety 37
The European Spirit (Jaspers) 45
existentialism 45
extreme utopianism 50

fanaticism 39; Russell's views on 16–18
fear 19
Federation of American Scientists 8
flexible response 35
force de frappe 37–38
Franck, James 7
freedom 52, 54
Freundlich, Elisabeth 67
future of man, Russell's views on 18–19
The Future of Mankind (Jaspers) 46–56

Gandhi, Mahatma 49, 51
geography, nuclear strategy 37
Germany 44; aftermath of World War II 46; totalitarianism 51
Goethe, Johann Wolfgang 79
graduated response 36
graduated retaliation 35–36

Has Man a Future? (Russell) 18–19
H-bomb, Russell's views on 11–12; *see also* atomic bombs
Heidegger, Martin 4
Henrich, Dieter 3
Hiroshima 2, 7, 11, 33, 44, 63–64, 67–68, 72, 75–78, 81
Hiroshima memorial 77
Hitler, Adolf 47, 58
Hobhouse, Leonard Trelawny 20
Hobson, John A. 20
Holocaust 80–81
human condition 48, 71–72; moral schizophrenia 70–71
human existence 50
human nature 19

Husserl, Edmund 4
Hutchins, Robert 8

imagination 37; Anders' view of 69, 71–72, 75–76
inadequacy of feeling 66
individualism 45
industrialization 29
injustice, Jaspers' view of 53
International Atomic Development Authority 9
International Authority 8, 10–11, 17, 19
international politics 39
international relations 4, 20, 26, 28, 31, 39
Israel, Jaspers' view of 53

Japan; *see also* Hiroshima; Nagasaki
Jaspers, Karl 3–4, 44–47, 72; aggression 53; atomic bombs 57–58; boundary situations 46, 57–58; change 54–55; democracy 55–56; equality 45; ethical idea 50–51; *The Future of Mankind* (Jaspers) 46–56; human condition 49–51; injustice 53; nonviolence 51; nuclear weapons 49–51; pacifism 44–45; politics 47–49; rational thinking 54–55; reason 50, 55–58; self-sacrifice 50–51; suprapolitics 46, 48–50; totalitarianism 51–54
Jungk, Robert 72, 73

Kafka, Franz 72
Keynes John Maynard 20
Khrushchev, Nikita 15, 18
Killian, James R. 12
killing, Anders' view of 80
"The Kind of Fear We sorely need" (Russell) 10
Korean War 1, 4, 28

Laski, Harold 20
Leghorn, Richard 34–35
Lilienthal, David 9

Lucky Dragon No. 5 11
Lukáks, György 45

Machtpolitik 39
The Man on the Bridge. Diary of Hiroshima and Nagasaki (*Der Mann auf der Brücke. Tagebuch aus Hiroshima und Nagasaki*), Anders 73–74, 76, 77
Man's Peril (Russell) 12, 14
Marshall Islands, Bikini test 11–12
Marxism 52, 56
massive retaliation doctrine 34–35
McNamara doctrine 35, 37
Metallurgical Laboratory (the Met Lab) 7
missile gap 35
Modern Man Is Obsolete (Cousins) 8
Mumford, Lewis 3
My Jewishness (*Mein Judentum*, Anders) 63–64

Nagasaki 2, 7, 33, 44, 63–64, 67, 76–77
natality 48
national belonging 10
nationalism, Russell's views on 16–18
Nazis 47
Nazism 63
negotiated peace 28
Nicomachean Ethics 64
non-being, Anders' view of 76–78
Nono, Luigi 82
nonviolence, Jaspers' view of 51
Nothing 76–78
nuclear disaster 80–81; *see also* Hiroshima; Nagasaki
nuclear energy 14–15
nuclear force de frappe 37–38
nuclear strategy 34; Aron's view of 34–38
nuclear tests 15; *see also* Bikini test
nuclear weapons 1–2; Anders' view of 73–76, 78–81; Aron's view of 29–30; Jaspers' view of 49–51; totalitarianism 51–54; *see also* atomic bombs

88 Index

Obstacles to World Government (Russell) 9
Of War (Aron) 27
Old Continent 35
One World (Willkie) 8
Open Letter to the United Nations (Bohr) 9
Oppenheimer, Robert 9
Outdatedness of Human Beings (Anders) 65, 70, 73

pacifism: Anders' view of 81; Jaspers' view of 44–45; realistic pacifism 80
Parliamentary Group for World Government 12
passivity 51
peace: Anders' view of 75; Aron's view of 28–30; negotiated peace 28; Russell's views on 10, 13–14
Peace and War (Aron) 26–27, 40
Peace and War among Nations (Aron) 32
peace process 13
Peace Tower 77
Plato 69
political action 48
politics, Jaspers' view of 47–49
Powell, Cecil 15
pre-emptive strike 33
Pre-emptive War 33
Preventative War 33
Promethean discrepancy 65, 70
prudence in international politics 39
psychic-emotional blindness 71
Pugwash Conference on Science and World Affairs 15

The Question of German Guilt (Jaspers) 46, 48, 56

racial antagonism 10
radioactive contamination 78
rational thinking, Jaspers' view of 54–55
realistic pacifism 80
reason 50; Jaspers' view of 55–58
reasonable response 36

reconstruction 77
Rencontres 45
"return of the repressed" 2–3
"The Road to Peace" (Russell) 13
Robbins, Lionel 20
Rotblat, Józef 8, 15
Russell, Bertrand 3, 20, 49; atomic bombs 7–8, 12–14; Campaign for Nuclear Disarmament 16; détente 14, 17; education 18; future of man 18–19; H-bomb 11–12; nationalism 16–18; need for change 8–9; overcoming fanaticism 16–18; peace 10, 13–14; Russell-Einstein Manifesto 14–15; war 6–7, 19; Western values 9–11; World Authority 13–14; world government 9–11
Russell-Einstein Manifesto 14–15
Russia 2–3, 9, 52–54

Sartre, Jean-Paul 75
schizophrenia 70–71
Schweitzer, Albert 69
scientific totalitarianism 11
self-defence 53, 80
self-sacrifice, Jaspers' view of 50–51
socio-economic cleavage 10
solidarity 46; age of deterrence 30–34
Soviet Union 48; age of deterrence 30–34; nuclear power 12; nuclear tests 15; scientific totalitarianism 11; totalitarianism 51; see also USSR
St. Laurent, Louis 14
"state of necessity" 78–79
statesman 55
supraliminal events 70
suprapolitics 46, 48–50

technocracy 65, 70
thermonuclear dual 36
Thirty Years' War 28
threat of nuclear war 32
totalitarianism 45, 51–54

Ukraine 2–3
UNAEC *see* United Nations Atomic Energy Commission (UNAEC)
United Nations 44; aggression 53
United Nations Atomic Energy Commission (UNAEC) 9
United States 53; age of deterrence 30–34; atomic bombs 9; H-bomb 11–12; nuclear strategy 34–36; nuclear tests 15; nuclear weapons 29–30
Usborne, Henry 8
USSR 2, 9, 17, 26, 29, 32–33

violence, Anders' view of 79–81
vulnerability of thermonuclear apparatus 32–33

war: Aron's view of 28–30; Russell's views on 6–7, 19; *see also* nuclear weapons
We, the Sons of Eichmann (*Wir Eichmannsöhne,* Anders) 71

Weber, Max 47
Weizsäcker, Carl Friedrich 73
the West 53; Jaspers' view of 52
Western values, Russell's views on 9–11
Which Way to Peace? (Russell) 6
whiteness 53
Willkie, Wendell Lewis 8
Wilson, Woodrow 6
Wootton, Barbara 20
World Authority 13; Russell's views on 13–14
World Federation of Scientific Workers 8
world government, Russell's views on 9–11
World Government weeks 8
World War I 6, 29, 47, 63, 76
World War II 7–8, 20, 26–27, 29, 44–46
"world without life" 64